How to save the world for free

With love and
happy fishes!
Natalie x

To Elliot, who I hope will grow old in a world where albatrosses still exist and they eat fish instead of plastic.

Published in 2019 by Laurence King Publishing Ltd
361–373 City Road
London EC1V 1LR
Tel: +44 20 7841 6900
Fax: +44 20 7841 6910
Email: enquiries@laurenceking.com

www.laurenceking.com

A catalogue record for this book is available from the British Library.

ISBN: 978-1-78627-499-1

Cover design: Lidka Wilkosz and Mariana Sameiro

Interior design: Nicolas Pauly and Manisha Patel

Proofreading: Emily Asquith

Printed in China

This book has been printed on FSC-certified paper and uses nontoxic vegetable-based inks.

Laurence King Publishing is committed to ethical and sustainable production. We are proud participants in The Book Chain Project ® bookchainproject.com

How to save the world for free

NATALIE FEE

Illustrations by Carissa Tanton

Laurence King Publishing

CONTENTS

Save the World When You...Travel 77

Save the World Where You...Live 94

Save the World When You...Bank 108

Save the World When You...Dress 118

Save the World When You...Use the Bathroom 130

Save the World When You...Bleed 141

Save the World When You...Exercise 149

Save the World When You...Have Sex 157

Save the World When You...Play 170

Save the World When You...Vote 180

Maximize Your Impact 191

INTRODUCTION

You're here. Somehow, because of an inkling that you need to play your part in saving the world, or through a series of serendipitous happenings that led you to be holding this book in your hands, or simply because you saw it lying there or had it flash up on your screen and were curious enough to pick it up or click on it, you've arrived at this moment, with me.

And I'd say it's quite a significant moment, this. Not just because we're meeting, probably for the first time, but because we're about to form an alliance. You, me and potentially everyone reading this book, who decides that saving the world, in whatever small or great way we choose, is quite possibly the most fantastic thing we could do with our time here on Earth. And not just because we might be doomed if we don't, but because we can and because it's going to be fun.

To say that we can face the realities of climate breakdown, rising sea levels, loss of biodiversity and ongoing pollution without descending into a spiral of despair may seem counterintuitive. And, sometimes we need to visit those feelings of disbelief, anger or despair in order to find the motivation to act. But despair in itself won't save the world. Beauty, on the other hand, might. Imagining a more beautiful world, creating community and being part of a groundswell of positive action is a much better option. And that's the kind of alliance I'm talking about.

So, this is how we're going to save the world. Oh – and we're going to do it for free, too. So that everyone can join us. Yes, a few suggestions in this book may come with an initial cost, such as investing in a good-quality safety razor instead of disposable ones, or switching to a menstrual cup instead of tampons or pads, but they will *all* save you money over time.

This journey – and it is a journey, because saving the world is a series of small steps along a winding path, as opposed to an overnight fix – is about you and your relationship to the world. I, for one, won't be judging how well you're doing, how fast you're going or how many followers are hanging off your every Instagram post. I will, however, be actively championing anyone who makes a change in their life and discovers that the alternative to what they were doing before is bringing them more joy. Whether it's finding that by cutting out processed meat you notice your health improving, or that by switching from the gym to a community running scheme you make a whole new bunch of friends, this saving-the-world business won't mean much to you unless it feels better than the way you used to do it. When saving the world makes you feel good, it's worth sharing.

Let's face it, we've got a whole heap of serious environmental issues on our hands and the Earth is starting to show signs of pressure. With a growing population and increasing demands on the Earth's finite resources, combined with erratic weather, water shortages, ocean pollution and melting polar ice caps, us humans need to make some big changes pretty fast if we want to thrive, together, over the coming decades. The alternative isn't pretty.

On the most basic level, saving the world is about protecting the air we breathe, the water we drink, the food we eat and the crazy-cool collection of wildlife living on Planet Earth, including us. And, on top of all that, this world gives birth to rainbows. I, for one, don't want to live in a world without rainbows. Or butterflies for that matter.

I can't ignore the blood on my hands either: children around the world born with birth defects as a result of the chemicals used to make the clothes we wear; seabirds starving to death with stomachs swollen with plastic used to wrap our food and drink; whales tortured and dying from being blasted with high-intensity sonar used by both the military and oil and gas corporations searching for fossil fuels to power our cars and homes. It's not ok. Everything is not fine as it is.

We're saving the world because we feel inspired – or compelled – to. Or, because we've reached a point when enough is enough. Forget about the people who aren't feeling any of these things for now. By focusing on you, by learning to channel your talents into Earth-positive ways of doing what you do, as an artist or an asset manager, or an asset manager who makes art in their spare time, you'll experience a deepening sense of meaning, purpose and wellbeing. And those qualities, by their very nature, ripple out and make waves.

If our motivation for saving the world stays rooted in fear, panic or anger, we risk not only harming ourselves but also shutting any windows that may have been open to the winds of change. Feel the feelings, let them shake you to your core and then channel them into positive action. And should you feel overwhelmed at the prospect of making so many changes in your life, just take it one step at a time, focusing first on what feels most doable. My guess is that those first small steps will feel so satisfying that you won't be able to stop yourself from taking another. And, maybe one more after that. Who knows where it'll take you?

I had no idea that watching heartbreaking, devastating scenes of albatross chicks dying with their bellies full of plastic on Facebook would lead to me devoting the next five years (and counting) of my life to doing whatever I could to stop this tragedy from happening. Or that it would lead to my setting up and running a multi-award-winning environmental campaigning organization.

It makes total sense to me to do good in the world. I learnt that, when I dedicated my time, energy and talent to creating and working on things that supported life on Earth, I felt better. When I went beyond myself and what I thought I wanted, towards what needed me most, I changed for the better. And that's my wish for you: that by doing some of the things in this book, the ones that make your heart sing, you'll feel a growing sense of wellbeing, creativity and connection.

How you use this book is entirely up to you. Start to finish, back to front, middle first or open at random. It's really that kind of book. Of course, it's not a bad idea to start at the beginning and familiarize yourself with the whys and wherefores of saving the Earth, but it's really not essential. The only thing you need to do is try some of the stuff suggested in these pages. I don't imagine many of you reading this will do everything (but, if you do, you're my hero, please write to me!) but lots of you will do some of the stuff and see which bits stick, which ones don't and which ones you actually love. Naturally, the ones you love will be the keepers and the things you become a great ambassador for.

Finally, this book is not designed to make you feel guilty, ashamed or not good enough. The very fact you're here and you're interested in saving the Earth is a wonderful thing.

I was tempted to assume that, given you're reading this book, you're well aware of the environmental issues that threaten life on Earth. But, in case you're not, or you think that we only have the problem of plastic pollution to solve, we'd better cover some ground. Some parched, over-fertilized ground at that.

If you are fully up to speed on the current challenges we face, and would rather not remind yourself of the breadth, imminence and magnitude of them, then please feel free to skip the next chapter and get going with the actions. When it comes to saving the world there is no greater ally than 'Now'.

One last thing. This book isn't what you would call a heavyweight when it comes to the nuts and bolts of climate change, dwindling resources, plastic pollution or any of the subjects listed below. I'm not an academic, as you may have already guessed – I tend to learn enough about something, or feel moved enough by something, to go and do what I can to address it. But that's just me and, thankfully, we're all different and you may decide that you want to learn more about the topics we're going to cover in this book.

Whilst experts and scientists have written many of the studies I refer to in this book, we don't all need to be experts to make change happen. If 100,000 readers aged around 25 decided never to buy plastic-bottled water again, that would be around 1.14 billion plastic bottles saved and around 86,000 tonnes of carbon-dioxide emissions stopped.[1] That's the same amount of carbon dioxide generated from powering over 10,000 homes for a year. And that's from taking just one simple action.

Despite being the problem, we're also the solution. Funny that. Our actions matter and every small step we take towards solving the problems we face counts. So, let's get down to it and begin our overview of the major environmental issues we're facing. And I invite you to breathe deeply while finding out that we're well and truly up shit creek – the good news is that we have enough paddles for everyone.

The Seven Things We Need to Save

1: OCEANS AND SEAS

PLANET OCEAN

With almost three quarters of the Earth's surface covered in salt water, it's pretty obvious to see that if the oceans are in trouble, the whole planet's in trouble. They don't just roll around doing nothing: our oceans provide us with 70 per cent of the oxygen we breathe, they regulate temperatures, and 3 billion of us rely on them as a primary food source. So, pretty significant for the existence of human life. And scientists and natural-history experts the world over agree that there's more than a commotion in the ocean – there's a crisis.

Ice is melting at unprecedented rates at the North and South Poles, carbon dioxide is making our seawater acidic, coral reefs are being bleached to death because water temperatures are rising, entire ecosystems are being destroyed by overfishing, oxygen-starved 'dead zones' are appearing as a result of fertilizer and sewage running off the land into the seas and, on top of that, the oceans are being filled with plastic.

THE PLASTIC PROBLEM

Since the first single-use plastic bag was patented in 1965, we've made over 8 billion tonnes of plastic.[1] And, while plastic has been a miracle for the healthcare industry, around half of it is used to produce single-use packaging. In the case of straws, coffee cups, lids and stirrers, we're using a material which lasts forever, to make things we'll use for only a few seconds. Plastic doesn't biodegrade (it fragments into minute particles called microplastics), and around 12 million tonnes of it ends up in the ocean every year from littering, leaks and plastic-pellet spills from the plastic-manufacturing industry, the fishing industry, our sewers and our storm drains.[2] That's the equivalent of a truckload of plastic entering the ocean every minute. According to the Monterey Bay Aquarium, in California, around a million seabirds and 100,000 sea animals die from eating, or being caught in, plastic every year, and microplastics are now found in the most remote parts of the oceans, in the seafood we eat, sea salt and in 90 per cent of table-salt brands.[3]

OVERFISHING

Sadly, there are not plenty more fish in the sea. Well, there would be, if we didn't keep eating them all faster than stocks can replenish.[4] And, as ever, it's not just what we do but how we're doing it that's the issue. Many fishing practices are unsustainable, not just in terms of depleting fish populations, but through the havoc wreaked on marine ecosystems through on marine ecosystems through by-catch – when seabirds, sharks, whales and dolphins are caught incidentally in fishing gear and killed – and from bottom trawlers, which are responsible for the destruction of entire habitats on the ocean floor. Overfishing.org say that almost 80 per cent of the world's fisheries are over-exploited, depleted or in a state of collapse. And, worldwide, about 90 per cent of the stocks of large predatory fish are already gone. It's serious stuff. The ecological balance of the oceans is under huge stress from overfishing and we know that if the top predators or keystone species (the ones that play a critical role in maintaining the structure of a specific habitat) are taken out of the equation, the food chain collapses. Like the 95 per cent of southern bluefin and Pacific bluefin tunas. Gone. And around 80 per cent of all the top predatory fish in coastal areas of the North Pacific and North Atlantic.[5] Gone. Buying 'dolphin-friendly' tuna isn't enough – we need a combination of protected marine-conservation zones, stronger international laws and more sustainable fishing practices, along with reducing how much fish we eat to solve this one.

HOT WATER

As if pollutants from industrial farming, melting ice, plastic pollution and overfishing weren't enough problems on the big blue plate, we're about to serve up another dollop or two of trouble. Our oceans are acidifying as a result of increased carbon dioxide in the atmosphere from our obsession with burning fossil fuels. You've probably heard that, in terms of human health, we need a healthy pH balance: not too acidic, not too alkaline. Well, the same could be said of the oceans. They're designed to absorb carbon dioxide, but too much of it makes them acidic. Ocean acidification has all kinds of negative consequences for marine ecosystems – it inhibits shell growth in marine animals like oysters[6] (which do an amazing job of filtering pollutants out of the water, so we could do with hanging on to them), it's suspected as a cause of reproductive disorders in some fish and, along with raising sea temperatures, is linked to coral bleaching. We really, really need our coral reefs too. These things can seem like far away, 'not my problem' issues, especially if you live inland. But, just as your body needs all its parts working together to function healthily, so the body of Earth needs all its parts working together too. We depend on Earth for everything and so, if something's wrong with the oceans, it's everyone's problem.

2: FORESTS

RAINFORESTS – 'JEWELS OF THE EARTH'

If oceans are the left lung of the Earth, rainforests are the right, generating around a fifth of the world's oxygen. They cover just 6 per cent of the planet's surface and yet they support around two thirds of plant species and a quarter of insect species.[7] Rainforests are like air-conditioners for the planet, helping to cool our warming atmosphere not just by absorbing vast amounts of carbon dioxide but also by soaking up sunlight and, through evaporation, creating clouds and rain. They're spectacular – despite our own spectacular efforts to raze them to the ground – still managing to provide ingredients for breakthrough medication and a home for 30 million species of animals and plants, as well as up to 84 uncontacted human tribes.[8]

FORESTS – A SPONGE FOR EMISSIONS

Forests all over the world, including temperate forests and rainforests, are a vital buffer against climate change. One recent study found that the world's forests absorb almost 40 per cent of the 38 billion tonnes of carbon dioxide created by humans every year.[9] Scientists have also discovered another fascinating way trees all over the world help to create clouds. They found that trees release gases into the atmosphere which is essential in a process called cloud seeding, in which water vapour clings onto these molecules to form a cloud.[10] And what do clouds do, other than look cool and rain on us? They bounce sunlight back into space, keeping us from overheating.

NOW YOU SEE IT

Despite the vital role trees play in protecting biodiversity and soil, in feeding and supporting healthy oceans and replenishing groundwater and rivers, governments around the world still allow corporations access to pristine, ancient forests to be logged. Except for Norway, which became the first country to ban deforestation, in 2018. More like that please. Rainforests are being cut down at the rate of around 30 million acres per year to provide hardwood for furniture and to clear land for agricultural use.[11] Around 80 per cent of the world's primary forests are gone, with almost 90 per cent of West African and Madagascan rainforest destroyed.[12] Let's just pause, take a breath, and say that again. Almost 90 per cent of West African and Madagascan rainforest has been destroyed.

And there's more to losing this lush and essential part of the biosphere than meets the eye. Rainforests are now so degraded that they are releasing more carbon annually than all the traffic in the US.[13] According to a recent report, tropical forest loss – through logging and clearing to turn into pasture – currently makes up around 8 per cent of the world's annual carbon-dioxide emissions.[14] Put in terms of rankings, if deforestation were a country it would be the world's third-biggest emitter, marginally lower than the US and quite a lot higher than Europe. And, when the land is turned into pasture for methane-belching cows, climate change gets a triple-whammy.

Protecting what remains of the rainforests and reforesting what we've lost is the simplest and cheapest way to stop climate change, but it seems our governments can't see the wood for the trees.

SOON YOU WON'T

At current deforestation rates, in ten years' time all the Indonesian rainforest will be gone, largely because of the palm-oil industry, and the same will be true for the rainforest in Papua New Guinea in around 13 to 16 years.[15] This is bad news for the warming planet and also for biodiversity. According to a 2014 study by scientists at the World Wide Fund for Nature (WWF) and the Zoological Society of London, the number of wild animals on Earth has halved in the past 40 years. More on biodiversity loss coming up.

3: THE ATMOSPHERE

ALL I NEED IS THE AIR THAT I BREATHE AND TO LOVE YOU

When Earth first formed, 4.5 billion years ago, it had no atmosphere – it was just a lump of molten rock spinning through space. Then, thanks to gases being spurted from volcanoes, followed by a long cooling-off period and the evolution of some very important organisms, oxygen levels soared and gave birth to the atmosphere as we know it.

Our atmosphere today is a complex safety shield that keeps us alive, not only by allowing us to breathe but also by protecting us from solar radiation and cosmic debris. It also keeps us warm by trapping just enough of the sun's heat waves bouncing off Earth's surface...for now, at least. Our atmosphere's layers include the troposphere, where most of our weather happens, and the stratosphere (the place where most jet planes fly) containing the ozone that protects us from ultraviolet light. Our troposphere is supposed to be composed of 78 per cent nitrogen, 21 per cent oxygen and just 0.04 per cent carbon dioxide, plus some tiny amounts of other gases, although, right now, thanks to the amount of greenhouse gases we humans are emitting, we're doing a stellar job of messing up this life-supporting ratio.

A LOT OF HOT AIR

The current biggest threat to our existence on this planet (that we know of) is climate change, since we're steaming towards a 'tipping point' beyond which there's no return. The heating of our atmosphere comes from the greenhouse gases (mainly carbon dioxide, methane and nitrous oxide) we've been spewing into the skies since we discovered the fossil fuels that powered the industrial revolution. Since then, coal, oil and gas have been fuelling pretty much everything we do to stay warm, fed, lit, entertained and mobile. But scientists the world over agree we've been threatening our very existence the whole time. If global warming passes 1.5 degrees Celsius (2.7 degrees Fahrenheit) above pre-industrial levels, they predict a 'calamitous risk to the health of the planet and populations' – and, deep breath – in 2016 global temperatures measured warming of 1.2°C (2.16°F). As it stands right now, experts agree that we're headed for a total rise of 3°C (5.4°F) by the end of the century.[16] Which makes me wonder, if 1.5°C (2.7°F) poses a calamitous risk, what does double that look like for life on Earth? Within the next 20 years, it's looking like the Arctic will be entirely ice-free in the summer months[17], making things another half a degree hotter again (because white ice reflects heat, and land and sea absorb it) and releasing mega amounts of methane into the atmosphere from the permafrost beneath it. Methane traps up to 100 times more heat in the atmosphere than carbon dioxide, so it's bad news not just for polar bears (the poster-animal of the melting poles) but for us humans too.

Back to carbon dioxide. Scientists, who measure the amount of carbon dioxide in our atmosphere (from the top of a volcano in Hawaii, if anyone is asking) in parts per million (or 'ppm'), have warned that we cannot go over 350ppm without dire consequences. Yet we've already reached 415.4ppm – a level not seen for 12 million years – and are on track to reach 500ppm within the next fifty years[18], at which point the warming will spiral exponentially. So, we could be not only in air that's 3 degrees warmer, but in very hot water, with mass destruction to rainforests, extreme weather patterns, devastating food shortages – with resulting mass migration – and stagnant seas giving off poisonous hydrogen sulfide gas.

A PLAN FOR THE PLANET

Scientists have been warning politicians about global warming for three decades: it's clearly laid out in thousands of peer-reviewed scientific papers. Continue to emit greenhouse gases, and the world will become toxic. And the signs are already here for them to see: erratic weather, drought, rising sea levels, wildfires, heatwaves, air pollution and climate migration. Yet most politicians, and the fossil-fuel industry, continue to champion finite, dirty energy instead of harnessing the wind, solar and tidal energy that could power the world's needs many times over without killing us all in the process. We can turn this around, but it has to happen fast, and we're all going to have to do our bit to make sure our governments and businesses are doing theirs.

A BAD AIR DAY

Just one other thing, while we're on the subject. Unfortunately, when we release gases and other particles into the atmosphere, they stay there and accumulate, or gather in water droplets and rain back down on us. Fuel emissions and particulates, pesticides, cleaning products, spray-on deodorants, volatile chemicals from paint and building materials don't just disappear after we use them, they enter our atmosphere and accumulate. Today, people living in areas of high air pollution are 20 per cent more likely to die of lung cancer than others[19], even if they've never touched a cigarette in their lives. Ozone-depleting chemicals, once thought to be a bad hangover from the 1980s along with big hair and neon clothes, are making a comeback, depleting the ozone that protects us and plants from DNA-destroying ultraviolet rays.[20]

4: RIVERS AND ICE

LIFEBLOOD OF THE LAND

If rainforests and oceans are the lungs of the earth, then rivers are its veins. Rivers pump fresh water containing vital minerals, nutrients and seeds around the land, they provide habitat for wildlife, feed the oceans and give us drinking water. But we don't have as much water on the planet as many of us, especially those with taps, seem to think. In fact, if all the world's water fitted in a gallon jug, only a tablespoon of it would be freshwater (the rest is saltwater). A tablespoon! And still most of us tend to take it for granted.

WATER SCARCITY

Back in 2015, the World Economic Forum listed water scarcity as the largest global risk in terms of impact over the next decade. From drought in the most productive farmlands to the hundreds of millions of people in the world without access to safe drinking water, water scarcity is affecting every continent and is predicted, unsurprisingly, to get worse as global temperatures rise. To put that in the context of climate change, if temperatures rise by 2°C (3.6°F) beyond the agreed, achievable cut-off point of 1.5°C (2.7°F), the numbers of people suffering from water shortages will be 50 per cent higher than at the already challenging 1.5°C (2.7°F).

According to the UN, global water demand is set to increase by 55 per cent by 2050, mostly from increased demands from the manufacturing industry – more people wanting more stuff. Yet, it's already projected that more than 40 per cent of the world's population will be living in areas of 'severe water stress' by 2050.[21] This, inevitably, means that the wealthiest will carry on having lots of stuff, while the poorest will go thirsty. And, as if we didn't already have enough reasons to switch to a green-energy supplier, power-plant cooling is responsible for a thirsty half of total freshwater usage in Europe and the US.

Let's revisit that analogy – the one where if all the world's water fitted in a gallon jug, only a tablespoon of it would be freshwater – and further blow our minds. Over two thirds of that tablespoon is actually frozen. Our ice caps and glaciers hold most of the world's freshwater – so, when it comes to watering our crops, livestock and providing drinking water for 8 billion people or so, we've got about a teaspoon.

FOOD: A THIRSTY BUSINESS

On a global scale, agriculture is the thirstiest business of them all, using 70 per cent of the world's freshwater resources.[22] In fact, around 50 per cent of an average American's water footprint comes from growing food. Only around 10 per cent is from personal use in the home – the rest is used for generating energy and manufacturing clothes and other goods.[23] But what we eat makes a difference, as all foods were not made equal when it comes to how much water they use in production. A meat-eater's water footprint is around 5,000 litres per day (10,500 pints), whereas vegetarians use between 1,000 and 2,000 litres a day (2000–4000 pints).[24] Vegans use even less (although watch out for almonds – 80 per cent of the global supply of these thirsty nuts are grown in California's drought-stricken Central Valley and use more water than Los Angeles and San Francisco combined). To put all this in terms of home use, it takes around 50 bathtubs of water to produce one steak[25], or, if you're saving water by having showers, then make that 180 showers' worth of water to make a single pound of beef. If you shower every other day, that's a year's worth of showers!

TROUBLED WATER

As well as relying on rivers and streams for our water supply, we pump it out of the ground. An unsurprising result? Groundwater levels are getting lower. More people are using more water; and soaring global temperatures and drought are also dramatically reducing water supplies in some areas.

Cape Town could be the first major city to actually run out of water. As in many parts of the world, the population keeps growing but the rainfall keeps slowing. In Cape Town, they're working on creating desalination plants to make seawater drinkable, along with groundwater-collection projects and water-recycling programmes to solve the problem...and it won't be too long before other cities are looking to do the same.

MURKY WATER

Despite depending on our rivers for drinking water and growing food, we pump a lot of crap into them. Literally as well as metaphorically. Around the world, we're dumping millions of pounds of garbage, trillions of gallons of untreated effluent from factories, sewage from our homes and storm water from roads into rivers every single year. In the US, 860 billion gallons of raw sewage are offloaded into rivers each year.[26] In some developing countries, the problem can be catastrophic for rivers: 2.9 billion litres of waste water from sewage, domestic and industrial sources enter the Ganges in India each day[27], resulting in the devastation of its aquatic ecosystem and causing widespread disease from waterborne illnesses.

So it's no wonder that around 780 million people worldwide have no access to clean water at all. Or that many large rivers are on the verge of dying. The Mississippi is sometimes called 'the colon of America', and not just because it's full of shit. Aside from sewage, it spews out agricultural pollutants (which are even more harmful) into the Gulf of Mexico, where this surge of nitrogen-based fertilizer has created a 'dead zone' in which ocean life has been suffocated from lack of oxygen.[28]

ICE IS NICE

It may not always feel like it, but we're actually currently living in an ice age, the Quaternary glaciation event, which began around 2.5 million years ago. We're in an 'interglacial' – or slightly warmer – period of that ice age, but we're in it nonetheless, since we have ice sheets at the Poles.

The world's vast reserves of ice are far away for most of us, but they have a majorly positive effect on all of our daily lives. For one thing, our polar ice keeps the planet at a liveable temperature by acting like a great cosmic mirror, reflecting the sun's hot, shining face back out into space. And the polar ice regions are home to some really quite magnificent animals that, ultimately, we'd all rather stayed alive and remain part of this thing called Earth.

I'M MEEEELTING

Polar ice is melting fast; the Arctic has lost 2 million square kilometres (772,000 square miles) of ice in 40 years, reaching record lows in the last few years, and latest studies show that

Antarctic ice is melting faster than ever before in recorded history.[29] In 2018, a ship sailed across the Arctic without the help of an icebreaker for the first time, and it is thought that ships will be able to sail across the North Pole by 2040, with some experts predicting ice-free summers there as early as 2020.

As the Arctic ice melts, it leaves areas of sea that absorb sunlight rather than reflecting it, as ice does, which is why warming is happening far more rapidly in the Arctic than elsewhere. The black soot that is dumped, in rain, on polar ice from all over the globe (that stuff we're breathing in too) adds greatly to this heat-absorbing effect. Temperatures in the Arctic are shocking scientists by reaching levels 20°C higher than usual and up to 33°C higher in the Russian Arctic (that's 36°F and 59.4°F, respectively). And, although retreating ice is normal in summer, it's the fact that this is happening during winter that's raising the alarm.

HIGH WATER

One of the challenges of connecting people with the consequences of climate change is that the effects are first being felt by those in hot and coastal areas. This means that most of us go on thinking it's not a problem. But rising sea levels could change that disconnect pretty quickly, when low-lying cities such as Miami, New York and London have to deal with flooding on a regular basis. And we can probably forget about beach holidays in the Mediterranean or the Maldives, because the beaches will all be under water – the rapid melting of the Greenland ice sheet is set to raise sea levels by 7 metres (almost 23 feet). The 'big melt' will also weaken the North Polar jet stream, which, in turn, will cause more extreme weather events – colder winters in the north and more severe droughts and floods everywhere. And the Gulf Stream, which brings the heat equivalent of a million power stations from the Tropics all the way up to the North Atlantic, doesn't like all this melted freshwater, either. According to the National Oceanography Centre in the UK, the Gulf Stream is already getting slower and, if it 'switches off', which experts predict will happen as the ice melts, Britain's winters will be getting a lot, lot colder.[30]

ENOUGH ABOUT POLAR BEARS ALREADY

Well, maybe just a bit more. It's not just polar bears that rely on ice as their home. Algae growing on the underside of sea ice is an essential food source at the bottom of the food chain. Zooplankton feed on it and they're then eaten by fish, which are eaten by seals, which are eaten by polar bears. And baleen whales eat zooplankton, too. The once impenetrable Arctic is disappearing, and the narwhals, foxes, harp seals, reindeer, caribou and others making up this special ecosystem are dangerously close to disaster. Oh, what an ice mess we've gotten ourselves into.

5: SOIL

SOIL IS NOT A DIRTY WORD

If the forests and oceans are the lungs of the planet and the rivers are its veins, then the soil has to be its skin. Soil anchors plant life to the ground and gives us over 95 per cent of our food.[31] And, while it may look like dirt to most of us, soil is positively bursting with life; there are more microorganisms in one teaspoon of soil than there are people on the planet. Healthy soil absorbs carbon, filters out impurities from water and fills the plants we eat with nutrients, which we then eat and benefit from. Soil rocks. But guess what? It, too, is under threat. But you probably figured that out, given it's made it onto the 'seven things we need to save' list.

FIFTY YEARS LEFT

Soil is one of our most important natural resources, right up there with the air we breathe and the water we drink. With very few exceptions, our food-production methods are exploiting the soil at an alarming rate, yet it's one of the least talked-about environmental issues. Scientists (where would we be without them?!) estimate that we only have 60 years of harvest left.[32] Half of the Earth's topsoil has been lost in the last 150 years because of our intensive farming methods. That's a full 30 football fields every *minute*.

WHERE DID ALL THE SOIL GO?

What, you mean the 24 billion tonnes of soil we're losing every year? Down the drain. Soil erosion is a major threat to our survival and is the result of the soil structure being disturbed. Natural vegetation (like a rainforest) is cut down to make way for growing food (like the soy grown to feed the cattle and chicken that end up on the menu at, among other places, McDonald's and Burger King).[33] Without decomposing leaves and vegetation filling it with nutrients, the soil quickly becomes infertile, so a new area is cleared, leaving the dead soil behind. Along comes the rain and washes the dead soil away. With the exception of organic farms, modern agriculture everywhere causes soil erosion, as does the overgrazing of the land. With no lush system of trees or vegetation to anchor it, the soil gets blown away by the wind or washed away by the rain and screws up other ecosystems, clogging up waterways and interfering with our already scarce drinking water supplies.

According to the UN, to keep up with growing food demands we're going to need an extra 14.8 million acres of new farmland every year. Yet 30 million acres of food-producing land is turning into desert annually as a result of soil erosion and degradation.[34] Even non-scientists can see that this doesn't add up; the way we're using our soil is completely unsustainable. It takes 1000 years to generate three centimetres of topsoil – we can't just put it back.

IT'S SO DEGRADING

As well as soil displacement, modern agriculture also causes a loss of nutrients from the soil. Unsurprisingly, yields are currently declining on 20 per cent of the world's croplands because of soil degradation.[35] Pesticides and industrial fertilizers give us short-term increases in the quantity of crops produced but, in the longer term, they weaken the soil. They disrupt the ecosystems and billions of microorganisms and native mycorrhizal fungi – remarkable and ancient soil organisms that have been benefitting plants for at least 500 million years – that were happily doing their thing until we blasted them with chemicals.

AN ENVIRONMENTAL HEADACHE

On top of everything else we're doing to it, soil is also getting whacked by a toxic mix of contaminants. Pollution from chemicals and radioactive materials from the rain, the air and landfill sites soak into the soil, even in some instances when we're trying to do good. According to German researchers, farmland can contain between 4 and 23 times more microplastics than the oceans.[36] Microfibres – tiny plastic filaments that shed from our nylon and polyester clothes when laundered – travel through the sewers to our water-treatment centres. Too small to filter out, they are either released into our rivers and seas with the treated water or mixed into the sludge from our poo...and spread onto our fields as fertilizer. Somebody make this chapter stop! Oh, wait, I'm writing it, and there's more to be said. We can't keep on giving soil less attention than it deserves.

IT'S AN ORGANIC MATTER

Soil also holds a key to combating climate change: it contains more carbon than the atmosphere and all the world's forests combined. And the healthier it is, the more carbon it holds. So, we need less tilling – all the mechanical digging, stirring and overturning that goes on in conventional farming and releases carbon dioxide – and more organic matter (decaying plant and animal material rich in nutrients and a key component of healthy soil). The UK's most excellent Soil Association claims that current UK regulation doesn't protect soil, and stresses that we should increase organic matter in soil by 20 per cent over the next 20 years to make it more fertile, increase yields, keep water in the soil to reduce flooding and sequester a load of carbon. Sounds sensible to me.

In fact, organic farms do all the key things the Soil Association recommends for saving our soils. Ready for the good news? Organic farming increases the amount of plant and animal matter going back into fields (which in turn increases soil organisms) by using manure and composts instead of nitrogen-based fertilizers; the non-cash 'cover crops', used between cash-crop cycles protect and improve the soil; planting more trees creates windbreaks for crops; soil compaction from machinery and livestock is reduced and helps the soil breathe more easily; and organic farming rotates crops, which helps prevent depletion of nutrients so that soil can recover more easily. I can almost hear the nematodes and earthworms wriggling for joy at the thought of it.

FISH FOOD

Some new technologies, like the use of mass greenhouses, hydroponics, aquaponics and vertical farms, are bringing some interesting food-growing technologies to the table. But, as it's early days, it is yet to be seen what the nutrient value of the foods grown with this technology would be. Ideally, we don't just want to get more calories from the same amount of land – we also need more nutrients from it. And some butterflies and bees buzzing around wouldn't go amiss either.

6: PLANTS AND ANIMALS

THE BIRDS AND THE BEES AND SOME KICK-ASS TREES

Our beautiful planet is teeming with millions of mind-boggling organisms – plants, animals, amphibians, birds, fish, bacteria, insects, fungi, reptiles, microbes and more. Octopuses with a brain in each tentacle, pink fairy armadillos, giraffes, narwhals and Venus fly traps. Dolphins, hummingbirds, seahorses, aye-ayes, gorillas and 400-year-old sharks (well, at least one of them). But it isn't just the miraculous fact of their sheer existence that makes these beings so very precious. Or that, in the case of the axolotls, just looking at one can make even the heaviest heart feel lighter. Nope, it's because we humans depend on them for our survival – even though *they* would be perfectly happy without *us*.

We have so many different species of plants and animals on Earth that, until a few years ago, we could only guess there were somewhere between 3 million and 100 million of them. Most species have never been described and many of them we'll never even know about, since we have a tendency to wipe out their habitats before we realize they are even there. So, when 'species counters' worked out a few years ago that there are about 8.7 million species, or a trillion if you count bacteria,[37] conservationists were very pleased that, at least, we now have a rough idea of what we need to save.

THE BASICS OF THE BIOSPHERE

The biosphere is the intricately woven balance of animals, plants and microbes that unites the world and atmosphere as we know it into one ecosystem. It's interlinked and intrinsic to life on Earth, and we humans (and millions of other species) have thrived because of it.

Here are some examples of this interconnectedness in action: apes keep rainforests regenerating by spreading plant seeds around the forest floor, which means more oxygen and less carbon in our atmosphere; termites chomp through plant matter so that we aren't drowning in piles of leaves; bats eat insect pests and pollinate fruit; ants aerate our soil. One thing depends on the other to survive.

And the bees? What, those little things? Why, they're only responsible for pollinating *70 of the 100 crop species that feed 90 per cent of the world.*[38] Plankton is the base of the whole food web in the oceans, which not only means we have majestic creatures such as blue whales, but also provides 3.5 billion people with a primary food source.

Our flora and fauna also heal us – where would we be

without aspirin from the white willow, blood-pressure drugs from the pit viper, or antibiotic penicillin? And, as microbes become drug-resistant, we'll more than likely need the natural reservoir of new, as-yet-undiscovered medicines of the future.

COMMITTING ECOCIDE

But, instead of looking after the biosphere that keeps our species fed and watered, most of us are blindly driving much of life towards a mass extinction. Not everyone, thankfully. There are millions of people that understand we're all connected and treat the Earth well. Weirdly, most of our politicians and big-business owners seemed to have missed that memo. Anyway, back to mass extinction. It's the sixth in the history of the planet, it's called the Holocene extinction and this time it's personal. As in, we've made it happen and it's happening really fast. According to the WWF, global populations of fish, birds, mammals, amphibians and reptiles declined by 58 per cent between 1970 and 2012. We've lost half the world's global wildlife population in just forty years.[39]

In 2018, a ground-breaking (or should that read 'heart-breaking'?) study showed that, since the rise of human civilization, 83 per cent of wild mammals and half the world's plants have been lost.[40]

If, at this point, you just want to put the book down and cry – or head out into nature to enjoy it while you still can – do it. Feeling grief is good. When we can truly feel the consequences of our actions we're more motivated to change them. But come back soon. Or just flick to the solutions section for some fast-acting inspiration for things you can do to make a difference.

So, where were we? Oh yes, we're still on mass extinction and now, specifically, 'insectageddon'. Studies on how to kill insects are far better funded than studies on how to save them, but one report shows that flying insects on nature reserves in Germany have declined by 76 per cent in 27 years.[41] No more insects, no more life. How can you tell how happy a cyclist is? From the number of bugs in her teeth. I like that joke. I'd like it to keep being true for generations of silly cyclists to come.

MORE THAN AN INSURANCE POLICY

Beyond thinking that we need to protect species 'in case we need them one day', we can choose to get a better understanding of how they all connect and how important they are to us. Killing off species as if there is no tomorrow puts tomorrow at risk – wildlife, water sources and vegetation all depend on each other for their survival, as do we. Rewilding, the reintroduction of top predators and keystone species into wild areas, can restore an entire ecosystem and is a great way to see our interdependence in action. This form of ecological restoration shows us how interconnected the health of the land and that of its inhabitants truly is. It also shows us the crucial part that animals we've previously vilified and labelled as dangerous play in our story. The reintroduction of wolves into Yellowstone Park, in the US, is a well-known example of rewilding, one simple change caused a 'trophic cascade' – a dramatic change in the ecosystem from the top down. The wolves ate some of the elks, taking the pressure off a number of plants, which slowed down streams meaning beavers could come back, and so on, until, over a few decades, an entire landscape was restored.

While we can't go around reintroducing wolves everywhere, rewilding projects that cause a trophic cascade illustrate how essential these keystone species truly are – whether that means playful sea otters regenerating kelp forests that sequester huge amounts of carbon dioxide[42], plankton at the bottom of the oceanic food chain or a super-cool shark at the top.

And we are in that mix, a keystone species blindly treating the biosphere as an expendable resource, risking our own lives without really meaning to. But it doesn't have to be that way. We can become 'conscious keystones', shifting our perspective of nature from something that provides us with interesting TV programmes and cut flowers for the table, to a deeper, more meaningful relationship that works both ways. The closer we get to nature, the more we love it. And the more we love it, the more we'll want to look after it.

7: US

SUPER HUMANS

Humans can do some clever things. We can fly to the moon, split atoms, perform heart surgery, make ice cream, and write operas and poetry thanks to our big brains and delicate digits. True, we also do a lot of horrific things, like killing and torturing each other, abusing sentient animals on a mass scale so that we can stuff our faces with toxic junk food and stockpiling lethal weapons in case we want to wipe each other out. But, beyond cleverness and cruelty, we're also reflective, perceptive beings who can learn from our mistakes, heal ourselves and be incredibly loving and funny at the same time.

TOO MUCH OF A GOOD THING

One of the things we're very good at is reproducing, while also lengthening our life spans using ingenious methods for keeping fed, watered, warm and well. This has resulted in population growth which, ironically, could be leading us towards extinction. D'oh! In around 12,000 years we've gone from being perhaps a few tens of thousands of hunter-gatherers dressed in organic, biodegradable animal skins, foraging for berries, making the odd kill and living in harmony with our environment, to being over 7.5 billion of us trashing the planet. In fact, we've trashed it so badly we've even created our own geological epoch, the Anthropocene, which, if dug up in a few million years' time, will be identified by all the plastic we squirrelled away in our landfill sites and sea beds. What a legacy.

You don't have to be a mathematician to work out that several billion people making babies means billions of babies, which is why the population is projected to reach 9.7 billion by 2050 and 11.2 billion by 2100.[43] The more people there are, the faster the population grows. Reaching the first 1 billion humans took most of human history, but the most recent increase of 1 billion was achieved in just over a decade. The glaringly obvious problem with this speedy rise in population is that we only have, wait for it, one planet! Experts say we have enough resources on the planet to feed and water around 10 billion of us, and that's if we all become vegetarians, since 10 billion vegetarians can be fed using the same amount of land as 2.5 billion omnivores, which means that, in the lifetimes of our current generation of children, we could quickly run out of food-growing space.[44]

IT'S NOT WHAT WE DO IT'S THE WAY THAT WE DO IT

To get some perspective on how vast the world is compared to how small we are, the entire global population could fit onto the island of Jamaica if we had 15 square feet each, as it is roughly 118 billion square feet (that's 1.39 square metres each). So, it's not that we don't have enough space, it's what we're doing to the space we have that's gotten us into hot water. As Gandhi famously said, 'The world has enough for everyone's need, but not enough for everyone's greed.' It could quite happily hold and feed all of us if we dialled back our desire for more things, as these things – on top of using up finite resources to mine, grow and make them – also release tons of heat-trapping greenhouse gases. Our obsession with getting more stuff is ultimately at the root of the problem, alongside our broken economic and political systems that got us acting this way in the first place. More on those later.

GETTING WASTED

We could start by sorting out a massive clanger: each year about a third of all food that's produced around the world is wasted. A shocking 35 per cent of all fish and seafood we haul out of the oceans never gets eaten, along with just under half of all fruit and vegetables.[45] We are creative; we can fix this. And we'll have to, seeing as how our planet's going to get a lot busier – and fast.

ALL THE LADIES

It's not all bad news though, as some experts predict the population could stabilize if global fertility rates continue to fall. Our population is exploding because better healthcare has led to lower mortality rates, but fertility rates have decreased globally by about half since 1960.[46] Women around the world are having on average just under 2.5 babies each. According to the UN, the sweet spot is an average of 2.1 children per woman: apparently, that's *replacement level*, the rate at which children replace their parents (and make up for people who die young).

Thankfully, there are solutions to overpopulation that don't require extreme measures such as the draconian one-child policy in China, which was enforced for 36 years until 2015: namely, access to a secondary-school education and safe, affordable family planning. Educated women have fewer children on average, are more likely to work full-time and make greater use of family-planning options. Studies in developing countries have shown that having a secondary-school education is one of the single most important factors in choosing to have fewer babies in later life; those who complete secondary school average at least one child fewer in their lifetime than girls who only went to primary school.[47]

In 2016, the UN Population Fund estimated that 350 million women in the poorest countries did not want their last child[48] but didn't have the means to prevent the pregnancy. Boosting women's rights through education and empowerment could change all this and turn our population crisis around.

DISENFRANCHISED, DISEMPOWERED AND DEPRESSED

Given how much stuff we've got, how much wealth we have on average and how much longer we're living, we don't seem to be very happy about it. In fact, from where I'm sitting, a large part of the population looks as if it needs a very long hug. Across the Global North (the richer, more developed countries of the world), most people are in debt, overweight and around a quarter of us suffer from mental-health problems.

And it's no wonder. Our electoral systems, which are supposed to be democratic, put power in the hands of an elite few instead of the many. Our financial systems are based on a currency without intrinsic value and the world economy is at risk of another financial meltdown. And our celebrity-obsessed consumer culture plays straight into the hands of the vast number of people who think they'll be happier if only they could have more stuff.

I've got a feeling that if we felt good about ourselves, like really, deeply good from the inside out, if we loved ourselves to bits, the world would be in much better shape than it is. So that's where we're going with this. Doing things to save the world has a positive feedback effect. We feel better about ourselves because we're making a difference and, because we feel better about ourselves, we feel better about other people too. We become creators instead of consumers, growing our own food, building relationships with people in our communities and coming up with ingenious new ways to make the most of what we've got.

ALL IS NOT LOST

Phew. After reading all that lot you may, rightly, be feeling quite glum. But I wouldn't be writing this book and you wouldn't be here reading it if everything was A-OK. So, at this point, before we move into the solutions, I want to say, 'Take heart!'

Some governments are starting to take action – but they still need pressure from you to do more and do it faster. Technology is changing – but it still needs innovators, investors and early adopters like you to make it work. And we are starting to wake up, in our millions, to the urgency of the situation. Plastic pollution is a brilliant example of how, when a story gets told enough times, people listen. 'Plastic is polluting our seas; we have to act now,' was the media's biggest environmental story in the UK and Europe in 2018. Pioneering NGOs (non-governmental organizations) and individuals were banging the drum, the media responded, masses then became aware of the problem and took action.

Then, when Sir David Attenborough talked about it on BBC's *Blue Planet II*, the government realized it could win votes by doing something about it, manufacturers and retailers realized they'd win customers by changing their product lines, and change happened. There's still a long way to go to stop plastic from entering our seas and a lot to be done in terms of creating a truly circular economy, whereby waste is designed out of the system. But the fact is, plastic pollution triggered a people-powered, grassroots movement that's changing the world for the better.

So, we can do this. In 2018, the Intergovernmental Panel on Climate Change said, if we want to limit global warming to 1.5°C (2.7°F), then we have 12 years to implement 'rapid, far-reaching and unprecedented changes'...basically, to sort our shit out. So, shall we get on with it?

Save the World When You...Eat

Every day you have three significant opportunities to vote for the planet, through what you eat for breakfast, lunch and dinner. The buying decisions you make every day are a simple, but powerful, form of direct action. What you eat, where it's from and what it's wrapped in account for a juicy 20 per cent of your carbon footprint, or 30 per cent if you include industrial agriculture's role in stripping the earth of trees.[1] The good news is that by making a few easy, simple switches you can make some satisfyingly chunky carbon reductions which, as well as helping to create a healthy future for the planet, are also scientifically proven to create a healthier future for you too. Let's get cooking.

COW BAD CAN IT BE?

Food production accounts for between one quarter and one third of greenhouse-gas emissions worldwide[2], and most of that is down to the meat and dairy industry.[3] In terms of reducing greenhouse-gas emissions, deforestation, water scarcity and ocean pollution, kicking the meat habit helps you tick all the boxes. Let's cut to the chase: meat-eaters are responsible for almost twice as many food-related greenhouse-gas emissions a day as vegetarians, and about two and a half times as many as vegans.[4] Which means that drastically reducing your meat and dairy consumption is one of the best things you can do to save the world.

Get your chops around this: the production of beef is around 10 times more damaging to the environment than any other form of livestock. Aside from beef needing 28 times more land to produce than pork or chicken, and 11 times more water,[5] the greenhouse gases emitted by cattle kick up a stink too. Cows are ruminants (that means they regurgitate their food, not that they're great thinkers), as are sheep and goats, which means they produce methane, a gas 23 times more powerful than carbon dioxide in its effect on heating up the atmosphere.

BE PART OF THE 'MOO-VEMENT'

A 2018 study conducted by comparethemarket.com showed that the number of people going vegan in the UK doubled from 2009–2016, growing to more than 3.5 million; the US is seeing a similar trend. A 2017 report showed that 6 per cent of US consumers now claim to be vegan, up from just 1 per cent in 2014.[6] That's around 19.5 million American vegans! Of course, people do this partly for health reasons, since red meat[7] with its antibiotic and hormone residues is likely to be a cause of disease.[8] But it's also the case that radically cutting back on red meat, choosing sustainably sourced fish, buying local and organic when you do eat meat and introducing other plant-based sources of protein, are some of the most powerful choices you can make if you want to make a difference to your personal health and the health of the planet. So get following some vegan hashtags on social media, get inspired and get creative with your quinoa.

GO ORGANIC

How we farm really does affect the quality of the food we eat. Organic means fewer pesticides, no artificial colours or preservatives, the highest standards of animal welfare, no routine use of antibiotics and, of course – no GM (Genetic Modification). Given the choice, would you rather eat a carrot that contains residues from pesticides and insecticides, or not? It's a no-brainer, isn't it? It's a yes to the one free from cancer-causing, sperm-zapping chemicals! Almost 300 pesticides can be used in non-organic farming and are often present in non-organic food despite washing and cooking. Research suggests that if all UK farming was organic, pesticide use would drop by 98 per cent![9]

And, again, given the choice, would you rather eat a carrot that has been grown in a field that has bees buzzing and birds singing in it, or a field devoid of life other than the people growing and picking it? Organic farms are havens for birds and bees – with up to 50 per cent more wildlife than non-organic farms. Another no-brainer, right? Yes, to the one that's been raised to the sound of music. One more question: if you had two carrots to choose from, and one tasted better (and contained more antioxidants) than the other, which one would you go for? The tasty one? You would, wouldn't you? Three fat juicy yeses to choosing organic carrots.

'BUT ORGANIC IS MORE EXPENSIVE'

Organic isn't just better for the planet and for animals – ground-breaking research published in the *British Journal of Nutrition* also found significant nutritional differences between food produced by organic and non-organic methods. This means, for every bite of organic food you eat you're getting significantly more bang for your buck. The research showed that organic crops can contain up to 68 per cent more key antioxidants, and that organic milk and meat contain around 50 per cent more beneficial omega-3 fatty acids than conventionally produced products.[10] I could leave it at that, as I think you're probably sold, except that this is a book about saving the world for free and buying organic can often be more expensive. Arguably, organic reflects the true cost of food production – industrial farming doesn't take in to account the cost of cleaning up the environmental mess it makes, nor chronic health issues and animal welfare catastrophes.

A PLEDGE FOR VEG

Growing crops organically not only generates less carbon dioxide than non-organic crops, but also helps mitigate climate change – healthy soils contain three times as much carbon as the atmosphere and five times as much as forests![11] In the UK alone, if all farming was converted to organic, at least 1.3 million tonnes of carbon dioxide would be taken up by the soil each year – the equivalent of taking nearly 1 million cars off the road![12]

LOSE THE PACKAGING

When it comes to food shopping, there are few things in the world more satisfying than getting to the till and seeing that nothing in your trolley is wrapped in plastic. There are several reasons why buying packaging-free food can help save the world. First of all, you're doing your bit to stop plastic from entering our oceans. Secondly, choosing loose produce creates less food waste, as you're able to select the perfect size and amount you need. Thirdly, it actually saves you money!

CHOOSE YOUR SUPERMARKET

Find whichever of your local supermarkets is best for loose produce and make that your first choice for food shopping. The more of us that shop in such places, the more the other supermarkets are likely to follow suit.

ZERO-WASTE SHOPS

Thankfully, zero-waste shops, which avoid single-use packaging, are on the rise. *And*, often they're cheaper than the supermarket alternatives. From pasta to porridge, cornflakes to coffee, you'll find an incredible range of loose produce to choose from. And, if you don't have a zero-waste store near you, well *there's* an invitation to become a shopkeeper if ever there was one. According to a poll commissioned in the UK in 2018 by the campaign Zero Waste Week, four out of five Britons said they were concerned about the amount of single-use plastics generated. A friend of mine who owns a health-food shop managed to crowdfund the money for the storage and dispenser units needed to sell loose products. The response from the local community was amazing and within three months her business went from surviving to thriving.

LOCAL MARKETS

The average meal you consume is extremely well travelled, clocking up 1,200km (745 miles) from farm to plate.[13] And, while transportation only makes up around 11 per cent of your food's carbon footprint, it sounds as if local means better when it comes to saving the world. But is it really?

Actually, the science of food miles is a bit complicated. Depending on how energy intensive your local farming methods are, it may be that produce imported from abroad has a lower overall carbon footprint if it has involved more traditional farming methods, with less dependence on artificial heat, and using greener local energy. Although this is rarely the case with air-freighted food, it can be the case with sea freight: 1,000 kilograms (2200 lbs) of air freight will emit around 500 grams (approx. 1lb) of carbon dioxide for every kilometre (approx. ⅔ mile) it flies on a modern cargo airplane, compared to 15 grams (approx. ½ oz) for every kilometre it sails.[14]

But it's not as if our supermarkets tell us if it's been shipped by air or by boat. So, what should we do? Get down to your local market, ideally a farmers' market, and talk to your producers. The 'best for you and best for the planet' option is to support and buy local, organic, seasonal produce. And, besides, a wicker basket hanging from your elbow, overflowing with greens and leaves and bunches of fresh-smelling produce, is a fine way to shop. There's nothing like meeting the people who've grown your food, or made the cheese themselves, and having a good old chat as you go.

BAKE IN BULK

One of the simplest, loveliest and tastiest ways to save the world is to cook at home more often. Forget about overpriced, over-processed sandwiches, or meal deals covered with plastic packaging, or spending a fortune every day on lunch. Nope, you can save money, time, energy and your health by cooking more than you need for supper and having the leftovers for lunch the next day. Sure, you can treat yourself to lunch from a local independent café once a week, but it's best if, for the most part, you can start cooking in bulk. You can stick extra portions in the freezer or take them to work and be the envy of your colleagues when you whip out a walnut roast with salad for lunch.

SHARE THE LOVE

According to the Food and Agriculture Organization of the United Nations, roughly one third of the food produced in the world for human consumption every year – approximately 1.3 billion tonnes – gets lost or wasted. That's the equivalent of an area larger than China being used to grow food that is never eaten. And, here's the worst bit, the world's nearly one billion hungry people could be fed *on less than a quarter* of the food that is wasted in the US, UK and Europe.[15] It's exasperating. So, we're going to need people like *you* championing the solutions, such as food-sharing and community-supported agriculture.

FOOD-SHARING SCHEMES

Food-sharing schemes are a great way to distribute your leftover fruit, veg and any other useable foodstuffs that you don't need. There are thousands of these initiatives around the world, all you need to do is find your local one and use it. And hooray to meeting more of the lovely humans in your neighbourhood! Search for 'food sharing schemes near me' or download the Olio food-sharing app for starters and get in the groove with sharing excess food.

COMMUNITY-SUPPORTED AGRICULTURE – FRESH, FRIENDLY AND PACKAGING-FREE

Picture this: it's the weekend. You fancy a stroll. You grab your reusable bag (the one with all the mud in it from last week's veg collection) and head out. On your way back home, you go via the 'shed', a conveniently placed drop-off point for you to collect your coming week's veg. You tap in the code to the lock, swing back the wooden doors and are greeted by stacks of fresh veg, a handwritten note from your veg grower telling you about this week's harvest, some news from the farm, a set of scales to weigh out your quota and a list to tick your name off. You fill your bag with seasonal produce and read that, in a couple of weeks' time, you're invited to a volunteer day at the farm, to help sow some seeds, dig some beds or help harvest a glut of squash. Just a few miles away, real people are growing real food, and you're eating it. It's generally free from any packaging, it's stuffed with vitamins and minerals and it has fewer food miles than anything – unless you've grown it yourself in your garden or on your windowsill.

Discovering my local Community-Supported Agriculture organization (CSA) was possibly the most transformational and exciting thing that happened to me when it came to changing the way I shop. I love it, and I think anyone would. A CSA gives you direct access to high-quality, fresh produce grown locally by regional farmers. All you have to do is become a member of your local CSA and sign up to buy a 'share' of vegetables from a local grower. That grower will then deliver your share of produce to a convenient drop-off location in your neighbourhood. But don't just take my word for it: type in 'CSA schemes [your town/city]' into Google or Ecosia (a search engine that plants trees!) and see what's growin' near you.

MORE WAYS TO SAVE THE WORLD WHEN YOU EAT

EAT A LOW-CARBON (NOT TO BE CONFUSED WITH 'LOW-CARB') DIET: lower-emission foods include pulses, quinoa, local and seasonal fresh produce.

ASK FOR A DOGGY BAG: restaurants are responsible for a large part of our food waste, largely because of oversized portions. Or search for your local 'zero-waste' restaurant; if you live in a large town there may well be one nearby.

USE OR COMPOST ANY LEFTOVERS: 'upcycle' rinds, peels, cores, stalks and bones into 'scrap stock', casseroles, soups or stir-fries. If your local waste collection doesn't include food, get a wormery in your yard for small-scale home-composting that won't attract rats.

ONLY BUY SUSTAINABLY CAUGHT FISH: beware of supermarket fish, especially tuna, even if certified 'sustainable' by the Marine Stewardship Council (MSC) as the practices of some of the fisheries they certify have been questioned.[16] Your local fishmonger will know the provenance of your fish, or look online for a local sustainable-seafood guide, such as the Marine Conservation Society's *Good Fish Guide* (UK) or Oceana's *Sustainable Seafood Guide* (US).

IF YOU DON'T HAVE A CSA NEAR YOU: order an organic veg box or produce box from your nearest food-delivery company; produce is often cheaper than from supermarkets, especially when there is a glut of a particular seasonal product.

USE YOUR OWN CONTAINERS: take them to the supermarket to fill at the deli counter, or baking parchment to wrap cheese or other goods in.

AVOID PALM OIL: as well as destroying orangutans' habitat, the stripping of tropical forests to create palm-oil plantations releases massive amounts of carbon dioxide into the atmosphere.

CHOOSE FOOD PACKAGED IN BPA-FREE TINS, GLASS AND PAPER OVER PLASTIC: BPA (bisphenol A) is the chemical used to make plastics and resins often found in packaging, especially tins, and has been proven to disrupt our own hormone balance.[17]

USE BEESWAX WRAPPERS INSTEAD OF CLING FILM OR ALUMINIUM FOIL: they are decorative and can be reused infinite times.

GROW YOUR OWN VEG: if you don't have a garden you could take on an allotment so you can cultivate the soil and produce free, nutrient-rich food, all while getting fresh air and exercise; you can also grow vegetables and herbs in windowboxes.

MAKE YOUR OWN ON-THE-GO SNACKS such as tamari-roasted nuts and seeds and do away with crisps and biscuits.

Save the World When You...Drink

We're all consumers these days. It's almost impossible to live without being one. A consumer is simply someone who buys goods and services for personal use – and who can avoid that? But an early definition of the word 'consumer' is 'one who squanders or wastes'. Take a look in the beverage section of the next shop you visit and see for yourself which of the two definitions feels more apt. You'll see a massive refrigerator whirring away, lighting up an array of juices, fizzy drinks, iced teas, coffees and herbal elixirs, all packaged up, that have been grown, harvested, processed and transported. At which point, you part with some cash, flip the lid, glug it down in a few minutes and look for your nearest recycling bin to throw the packaging away. It's thirsty work, this beverage business. So, my fellow squanderers and wasters, it's time to think about what we drink.

HOLD THE STRAW

You may have noticed that, these days, straws are increasingly hard to find in cafés, bars and restaurants, thanks to the efforts of the plastic-pollution movement worldwide. It's a fantastic example of how people-powered action can bring about positive change in the world. At giant chains, such as the pub brand Wetherspoons (UK) and Starbucks (globally), the last couple of years have seen a dramatic shift away from plastic straws towards paper ones, or just no straws at all. If you're not already in the 'no straw for me please' camp, try a search for 'turtle straw video' on the web. Seeing the horrific effects straws can have on turtles must be the surest way to get you to kick the plastic straw habit and, if you share it, possibly inspire a few million other people to do the same.

500 MILLION STRAWS A DAY

In early 2018, Americans were getting through nearly two straws a day. Each. That's around 500 million pieces of plastic, being used for about twenty minutes, ending up in landfill or escaping the waste system and ending up in the oceans, every day.[1] And then, taking a few hundred years, or more, to break down into microplastics.

IT'S JUST ONE STRAW...

I had a (minor) showdown in McDonalds recently with my son, who has yet to fully embrace the idea of our power as consumers and is rather partial to a milkshake. I encouraged him not to ask for a straw but instead to try using his mouth. 'But Mum, what difference is it going to make if I don't ask for a straw?' 'Ah, grasshopper,' I replied in my annoying way, 'So said seven and a half billion people.'

Some people with disabilities rely on straws to drink liquids, so encouraging a switch to paper straws keeps things more accessible. One alternative is to encourage retailers to provide straws only on request and for people who need a straw for medical reasons (or people who really like juices but don't want an orangey-green moustache) to carry a reusable – metal or glass – straw with them.

GET IN THERE FIRST

We've still a way to go when it comes to kicking the straw habit around the world. So, next time you're out for a milkshake, smoothie or gin and tonic, make sure to get your order in correctly *before* they slip a length of planet-polluting plastic into your tipple. 'Make mine a double...and hold the straw.'

'BUT IT'S BIOPLASTIC'

A word about bioplastics. Many restaurants and bars now say, 'It's ok, our straws are compostable!' In the move away from regular plastic, bioplastics are often touted as a solution. But, sadly, just because something is called 'bioplastic' doesn't mean it will break down in the marine environment. Bioplastics may have their place in the future but, for now, most municipal waste-management systems are not set up to recover and compost bioplastics, so they end up either being incinerated or going to landfill where they release methane as they break down – which is not good for climate change.[2] Very few products on the market are home-compostable, but new products are coming out all the time so do your research and, in the meantime, remember, *if you can't reuse it, refuse it.*

GET YOUR FILL

This one couldn't be easier. Of all the ways to save the world for free in this book, carrying a reusable water bottle has to be in the top ten. And you'll be in good company. The reusable-bottle market is soaring and, these days, it's almost more common to see someone swigging water from something stylish whether on the subway, in the park, or in the gym than from a single-use plastic bottle. To illustrate this happy story, according to a report by Transparency Market Research in 2016, the reusable water-bottle industry was valued at $7.6 billion in the US and it is currently expected to rise to more than $10 billion by 2024.

BOTTLED WATER – AN ENVIRONMENTAL CATASTROPHE

We're waking up, in our millions, from the stories we've been fed for decades from the bottled-water industry, who led us to believe that their product, which is often 500 times more expensive than the water flowing from your taps, is superior. It was quite possibly one of the biggest scams of the twentieth century. And it has been a key contributor to one of the ugliest environmental catastrophes of our time. For those of you reading this in parts of the world that tap water really isn't that great, we'll take a look at home filters in a bit.

According to data from organizations specializing in 'beach clean' data, such as non-profits #2minutebeachclean and the Marine Conservation Society, plastic bottles and bottle tops are consistently in the top ten items found on beaches in the UK, a finding that is echoed globally. Plastic bottle tops float in water, which is why they're more commonly found on beach cleans than plastic bottles and, as they're small, they also pass through storm drains (the metal grids on sides of roads designed to divert rainfall into our rivers and seas).

WHAT THE FRACK?

Legislation would ensure that bottle tops have to stay attached to their bottles during and after use. And deposit schemes, which ensure plastic bottles stay in the waste system and get recycled, would stop them ending up as litter or ocean plastic pollution. But, given that fracked gas from the US is already funding a massive boom in virgin plastics infrastructure across the world, we're better off ditching single-use plastics altogether and opting for reusables.

JOIN THE REFILL CAMPAIGN AND REFILL FOR FREE

It was the sight of plastic bottle caps inside the bellies of dead albatross chicks, captured so poignantly by American artist and photographer Chris Jordan, that inspired me to start a refill revolution. The Refill campaign, run by City to Sea, is now Europe's biggest free tap-water refill scheme. Shops, cafés and restaurants display a sticker in the window welcoming you in for a free top-up of your water bottle and register themselves on the free Refill app. Over 100,000 people have downloaded the app and are refilling for free, on the go, wherever they go.[3] If every Refill station on the app refilled just one bottle a day, the world would have 6.2 million fewer plastic bottles in circulation every year! No shops or cafés near you on the app? Head out into your community and start signing them up!

FILTER IT

If you really don't like the taste of tap water where you live, or you live somewhere where the tap water isn't safe, get yourself a water filter. Water filters remove impurities such as sediment and bacteria from water, and some remove heavy metals and fluoride. Some people say they also improve the taste, and some filters even claim to provide healthier water than both tap and bottled water. The more expensive they are, the better quality they are likely to be, but if you are swapping from bottled water to a filter you'll still save money in the long run and you'll be happier and healthier for it.

GET BUSY WITH THE FIZZY

If you're a lover of fizzy water and you're still buying it in plastic bottles, I have a money-saving, planet-friendly solution for you: get a Sodastream. The canisters are refillable and, if you're currently getting through around 4 litres (about 7 pints) of shop-bought sparkling water a week, you'll be saving money after the initial outlay for the machine after about eight weeks. Prefer smaller bubbles? Experiment with adding some still tap or filter water into your homemade fizz to get it just the way you like it.

CHOOSE TO REUSE

There really is no excuse for single-use when it comes to your coffee. Most major chains offer discounts if you bring your own cup and there are literally hundreds of ingenious and stylish reusable cups on the market now. From foldable ones to vacuum-sealed ones, glass to bamboo, it's really just a matter of working out your needs and getting yourself the perfect cup. My brother and his wife love the foldable Stojo cups I bought them, as they like to drink a coffee on their dog walks but don't want to carry bulky things around with them. Style-wise, check out KeepCup, Frank Green and Chilly's.

In the UK, we guzzle our way through more than 7 million disposable coffee cups each day[4] and, due to bad design and insufficient waste-management systems, they're not recycled. In the US, the situation is even worse, with around 25 billion polystyrene cups being thrown away each year.[5] But you could just ask yourself this: Do I really need to drink a coffee on the go? Can I get up five minutes earlier and have one at home? Can it wait until I get back to the office?

No? Ok, well then get yourself a reusable cup and repeat after me: if you can't reuse it, refuse it.

KETTLE TIPS

Ok, so this section may be more relevant to tea-guzzlers than coffee drinkers. Americans don't have the same levels of tea obsession as Brits and they tend to use stove-top kettles instead of electric ones, so this section is primarily aimed at the three quarters of British households who are over-filling their kettle when boiling water, wasting £68 million (that would be about US$88 million) each year![6]

Sitting down for a nice cuppa is part of British culture, but overfilling your kettle 24 times a week is a massive waste of electricity and money, as well as generating avoidable carbon dioxide; boiling more water than you need can easily add 20g (about ¾ oz) to the carbon footprint of your hot drink.[7] Do yourself, and the planet, a favour by keeping your kettle near the sink if possible and filling it up with just the amount you need, and no more. Keeping your kettle clean also keeps it efficient, since the more limescale it collects, the more energy it takes to boil.

MILK, MAN

While I don't drink dairy milk, I order two pints a week for my 15-year-old son who refuses to drink plant-based alternatives. Admittedly, I do make a point of proudly clinking my glass bottles when I bring them inside. Occasionally, a passing neighbour is beguiled by the sounds and falls hook, line and sinker for my routine. 'Do you get those delivered?!' they ask. 'Oh, you mean these?' I reply innocently. 'Yes! They're organic too. They cost a bit more, but it tastes better and means I'm cutting down on plastic.' They're sold. The nostalgia of milk deliveries in glass bottles and the urge to press down on that tight, silver-foil top overwhelms them to the point that they rush home, whip out their laptop, search 'milk deliveries in glass near me', fill out their details and breathe a sigh of happy relief.

Milk deliveries in reusable glass bottles almost became extinct, but they are now on the rise in the UK and US, thanks in part to the BBC television series, *Blue Planet II* (and the 'Attenborough effect'), and the plastic-pollution movement in general. Yes, milk in glass bottles is more expensive than its plastic counterpart. But here's a tip: some milk delivery companies offer £10 (around US$13 or US$14) off your bill every time you introduce someone to them. So, if you only use two pints of milk a week, by introducing one neighbour a month to the joys of milk in glass bottles, you could get your milk for free, at the same time as reducing carbon emissions by having more deliveries on the same street. Boom.

DRINK LESS MILK

My other cunning tip is to simply drink less milk. As we saw on p. 51, the dairy industry is the food industry's fourth largest emitter worldwide of greenhouse gases after beef, lamb or mutton, and farmed crustaceans, so you could consider going from eight pints a week down to four. The much-touted health benefits of cow's milk are questionable in any case; studies have shown that milk does not necessarily improve bone density and may also lead to other health problems.[8]

A PINT OF 'MYLK', PLEASE

Of course, you could do away with dairy altogether and switch to a plant-based alternative, such as one of the many grain or nut 'mylks' now available. I've only heard of one community scheme – and that's in London – that delivers it in glass. Now there's an action you could take, if this is something you're passionate about. Send a sad face to one of the big milk-and-grocery delivery companies, letting them know your plant-based, plastic-free milk happiness depends on them. Because, let's face it, Tetra Paks — which, even if they are recycled, will never become a Tetra Pak again — are rubbish for the planet too.[9]

MORE WAYS TO SAVE THE WORLD WHEN YOU DRINK

BUY LOOSE TEA AND COFFEE: fill your own bag at a shop which supplies loose tea and coffee and, if you use coffee pods, make sure to get biodegradable ones or buy a reusable one in stainless steel.

MAKE FRESH HERB TEA FROM THE GARDEN: make your own herb tea using mint, lemon verbena, sage or other herbs which are easy to grow in the garden or in a pot on your windowsill; it will taste so good and look so pretty, and costs little, if anything. If you don't manage to do that, choose organic herb teas in non-plastic tea bags.

WINE – RED, WHITE AND GREEN: buy wine and beer made in a country (or town!) as near to you as possible to lower the carbon footprint. And buy organic whenever possible as grapes are heavily sprayed with pesticides (plus, without the chemicals, the hangovers aren't as bad). Or ferment your own: there are no transport emissions and the heat produced will help warm your house.

MAKE YOUR OWN JUICES AND SMOOTHIES: making your own juices or smoothies means they will be fresh and nutrient-rich; it is cheaper than buying them from the shop, and means – yay – no plastic.

OAT MILK, NOT NUT MILK: if you go dairy-free, oat milk, which is easy and cheap to make, is one of the best options in terms of its carbon footprint. Avoid almond milk, or use it only sparingly, since almonds, which are mostly grown in drought-stricken California, use millions of litres of water to produce. Almonds also require the importation of large numbers of bee colonies which threaten native populations[10] and can, in any case, collapse themselves, most likely as a result of high insecticide levels in the state.

MAKE YOUR OWN KEFIR AND KOMBUCHA: these are fermented probiotic drinks which you can buy – at a price (usually in small glass bottles) – but they are easy and cheap to make using just a few ingredients, including a starter culture, and they're excellent for health.

SUPPORT A DEPOSIT-RETURN SCHEME: these pretty much do away with litter from plastic bottles and increase recycling rates (by up to 97% in some countries)[11] by providing an incentive to drinks buyers to return empty drinks containers in exchange for cash.

DITCH THE VENDING MACHINE: if your office has a vending machine, speak to your employer about getting them to ditch single-use plastic water and coffee cups and, ideally, switch to reusables.

CUT UP PLASTIC RINGS FROM PACKS OF BEER: these can end up in the sea and entangle marine life.

USE COFFEE GROUNDS IN THE HOUSE AND GARDEN: used coffee grounds can be used as natural pesticides, fertilizer for acid-loving plants, and compost. They can also absorb smells if you place dried grounds in an aerated tub somewhere you need a deodorizer (at the back of the fridge, for example, or in your car).

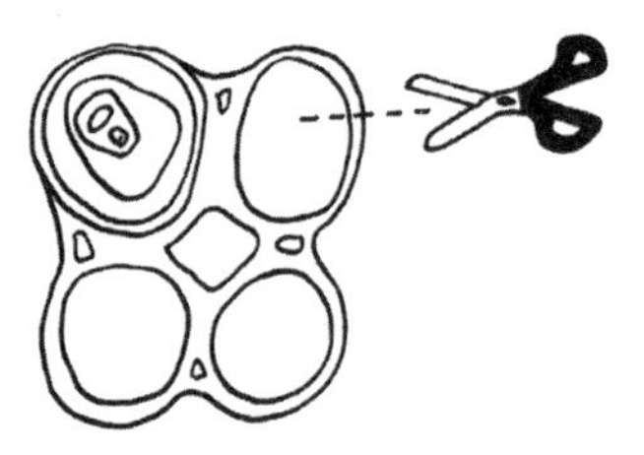

Save the World When You...Travel

We humans love to get around. There are people to meet and mountains to greet, temples to tiptoe around, parties to go to and, obviously, work to do that's not a handy half-hour cycle ride away. But, given that the world's transportation sector is responsible for 14 per cent of global greenhouse gas emissions,[1] and that even the average motorist in the UK and the US uses vast amounts of fossil fuel over their lifetime, we need to do some rethinking about how we get from A to B, how often we do it and what we do when we're there. So, quite a lot really. Let's break it down into bitesize little nuggets of doable things – free things at that.

CYCLING IS COOL...AND HOT

Cycling can save the world while making you fitter, happier and, wait for it, more attractive to people you might want to have sex with. Yep, it's official, not only is cycling cool, it's hot too. A study commissioned by the British Heart Foundation found that, out of 600 people surveyed, respondents believed cyclists to be 13 per cent more intelligent and cooler, and 10 per cent more charitable than other people; almost a quarter of them said they'd rather date a cyclist than another type of athlete.[2] Each time you hop on a bike and get a bit sweaty, you can throw into that heady mix an increase in endocannabinoids (uh huh, the same family of chemicals you'd be entertaining if you smoked a spliff – only better for you, because they were produced naturally, inside you), plus the fact that cycling increases your sex drive. So, you're left with an irresistible, irrefutable case for a great ride. As it were.

Who thought saving the world could be so damn hot? And, beyond getting your lycra-clad rocks off, cycling has so many other benefits. There are too many of them to list here, so I'll just choose my favourites to share with you. (Other than the one above. I think I made that clear.)

GET THERE FASTER

It costs less to go by bike and, if you're travelling through a city at rush hour, cycling will get you there faster. My commute to work takes about 35 minutes in rush hour by car, compared to 15 minutes by bike. So that's a no-brainer for me. (And for my butt, which stays in shape as a result.)

GET THERE CHEAPER

I spend under £50 a year (approximately US$65) on bike maintenance. If you're an employee in the UK and use your own bicycle for work (so, for cycling on business, not to and from work) you're entitled to 20p per mile (about 25 cents), tax-free. And, if you're self-employed, you can claim back the same on your expenses, plus you can deduct all the maintenance expenses. Huzzah! If you're pedal-powered in the US, check out The Bicycle Commuter Tax Benefit. The League of American Bicyclists, a non-profit organization, have a fantastic advocacy section on their website that can help you get the most out of your daily commute.

BEAT THE QUEUES AND SMELL THE FLOWERS

I'll never forget the happy (and slightly smug) free-wheelin' feeling I had when I first took a bike over to France. I left the car at the ferry port (looking back, I suppose I could have taken the train that far if I'd wanted to go even greener) and cycled onto the ferry for a fraction of the cost of people taking their cars. Being ushered off the boat first, out of the cloud of particulates (carbon monoxide and other toxic pollutants that were rapidly filling up the car decks) and into the fresh French air, was a buzz. In fact, the whole cycling holiday – my first – was a buzz. I remember cycling along and having a particularly ecstatic moment realizing I was travelling at the same speed as a bee, who was flying alongside me.

PEDAL POWER MEANS BRAIN POWER

Need more reasons to become a pedal pusher? How about because it makes you smart and can keep Alzheimers at bay as you age? A 5 per cent improvement in cardio-respiratory fitness from cycling leads to an improvement of up to 15 per cent in mental tests, according to researchers from the University of Illinois (who are all probably cyclists, as they're really clever).[3] They showed that cycling

helps build new brain cells in the hippocampus – the part of your brain responsible for memory, which deteriorates from the age of 30. Make a note of that bit before you forget. Or better still, go for a bike ride, baby!

CYCLING CAN SAVE THE WORLD

I hope I've inspired your mind; now, please allow me to blow it. According to the International Energy Agency, the car industry is responsible for roughly 15 per cent of world carbon-dioxide emissions. In the UK, over half of all car journeys could be made on a bike in less than 20 minutes.[4] More than half of all car journeys in London, for instance, are less than five miles. Imagine those journeys being made by bike. Half the pollution. Half the traffic. Half the time. And, what's more, 20 bicycles can park in the same space used to park one car. It takes a mere five per cent of the materials and energy used to make a car to build a bike, and, if just five minutes of the average 36 minutes a day people spend in cars in the UK were spent on cycling, the NHS (National Health Service) would see a five per cent fall in inactivity-related illnesses such as heart disease, diabetes and strokes, saving – wait for it – £250 *million* pounds a year (over US$325 million).[5]

So, let's go forth and cycle hard, cycle safe and cycle happy, knowing we're helping to save the world with every ride we go on.

FROM A TO B MORE JOYFULLY

Until our cities are covered in comprehensive, segregated cycleways (bring it on!) it stands to reason that for now, a lot of people will still be hopping in their cars. So let's take a look at some more of the greener – and cheaper – ways to get from A to B and back again.

CAR POOLS

When work is just that bit too far to cycle and public transport just a little too inconvenient, it's time to link up with a car-pooling scheme. Car-pooling, car-sharing, or lift-sharing – essentially sharing your ride – is by far the most fun way to get about if a car is your only option. You could even get there, on average, twice as fast, as many cities have HOV lanes (which is short for High Occupancy Vehicle, not High Octane Vroom as someone I know calls them); you get to meet people as extraordinary, or even more so, than you are and you save money. According to Liftshare, the UK's biggest lift-sharing scheme, the average commuter who shares a car could save over £900 a year (that's almost US$1,200)! And, in the US, Carbonfund.org estimate that a once-a-week shift to a car-pool can reduce your commuter carbon footprint by 20 per cent. Bye-bye road rage, hello good times, new friends and clean air.

CAR SHARES

Then there's the other type of car-sharing: car clubs. The kind when you offer out your car for others to use, and get paid for it, or borrow someone else's. Or you could join a city car club as an alternative to owning or renting a car. Car ownership is becoming old school. The sharing economy is booming and the icing on the cake is that you will always have somewhere to park, as car clubs have dedicated parking spaces. Perfect for day trips, weekend 'big shops', trips to the vet and picking up some second-hand furniture you nabbed on Gumtree or Craigslist.

ELECTRIC-CAR HIRE

Another option, if you're hiring a car, is to go electric. Electric vehicles are surging in popularity these days, partly because of concern over greenhouse-gas emissions and partly because they're cheap to run. They're thought to be far better for our air quality than traditional combustion-engine cars, since they emit minimal tailpipe pollutants, such as particulates and carbon monoxide. On the other hand, they're considered to have a 'long tailpipe', meaning that some of the polluting effects of the car are shifted elsewhere, including to the battery-production plants and the source of electricity. And electric cars, along with all other road vehicles, are still responsible for the number one source of microplastics in our ocean. Tyre dust, created by the friction between the road surface and the tyres, produces huge amounts of toxic dust and particulates which are blown into the air and washed directly into our rivers and seas each time it rains. Another reason to walk, go by bike or hop on a train.

WORKING FROM HOME

You will need to get your employer on board with this one, but a quick internet search of 'benefits of working from home one day a week' will give you more than enough ammo to blow their (bamboo, I hope) socks off. Studies have shown that working from home one day a week increases productivity and improves staff satisfaction.[6]

When you don't have to commute, you have more time to get down to your to-do list. You're less distracted by your colleagues who, although they may be brilliant, can also nibble away at your concentration levels with stories of their girlfriend's recent chocolate-brownie disaster. And you feel good that your employer values you and trusts you to get the job done from the comfort of your own home. In your underwear, if you so choose.

A report by the Carbon Trust says that, by working from home two days a week for a year, an average UK employee can shave off just under 4 per cent of their personal carbon footprint, save 50 hours' commuting time and also save £450 per year in travel (almost US$600) and other costs! And, avoiding unnecessary travel can be extended to the office too, by switching non-essential face-to-face meetings for a phone call, video conference or other way of telecommunicating with your business partners. Studies show that video conferencing can prevent millions of tonnes of carbon emissions in the US and UK alone.[7]

One more thing. Working from home is good for the oceans too, as you are more likely to whip up some leftovers for lunch and make your own coffee instead of buying on-the-go, over-packaged food and drink. Enough said.

PUBLIC TRANSPORT – A BREATH OF FRESH AIR

Why should we use public transport? Buses and trains are often late, they're frequently crowded, they're sometimes dirty, they don't go from door to door, and you can't cram the boot full of everything you might need.

On the other hand, we all know it would be better for the planet if we used public transport because, if we made the switch, vehicle emissions would be drastically reduced and we'd all be using less gas and breathing cleaner air. A full bus can take 50 cars off the road; a full train several hundred. A recent study showed that a switch to public transport could reduce emissions by half by 2050 and save 1.4 million premature deaths worldwide every year.[8] And, if you go by bus or train instead of car, you can reduce your household carbon footprint by a whopping 30 per cent.[9]

TRAINS AND BUSES ACTUALLY HELP YOU GET FIT

Here's what's in it for us, other than saving money and leaving some oxygen in the atmosphere for our children: using public transport can make us slimmer, happier and healthier.[10] A large study of 18,000 people showed that those who swap their cars for public transport are happier and sleep better. The researchers pointed out that using public transport is more relaxing, which makes sense when you think how much nicer it is to sit on a train or a bus reading a novel or learning Mandarin, or just looking out of the window, instead of sitting in a traffic jam or having other drivers angrily shaking their fists at us for no good reason. The health-promoting effect was put down to the exercise of walking to and from the bus stop or train station, and a US study showed that, for the same reason, people lose, on average, just over 3kg (7lbs) in a year by switching to public transport.[11]

WE WANT BETTER PUBLIC TRANSPORT!

In spite of its obvious advantages, public transport use in the UK has declined by 11 per cent since 2008 and, in the US, by 7 per cent.[12] That's why campaigners are asking governments to start creating better public-transport systems, so, if you think politicians who could make a difference are suffering from inertia in this area, you might like to find out who your local pressure groups are and add your voice. As congestion from traffic grows, fuel becomes scarcer, and carbon-dioxide emissions become even more of a problem than they are now, public transport could and should be a big part of the solution for the commuters of the future.

A JETSET-OFFSET MINDSET

If you regularly travel by plane, then the most effective way to reduce your carbon footprint – other than giving up meat – is to fly less often, if at all. If the aviation sector were thought of as a country, it would be the seventh biggest polluter in the world.[13] The even scarier news is that it's an industry set to almost double in less than twenty years. The International Air Transport Association (IATA) expects 7.2 billion passengers to travel in 2035, compared to 3.8 billion in 2016. But this is where the 'saving the world for free' bit gets difficult because offsetting costs money. I would usually recommend you offset any essential flights through a quality-assured scheme. And after lots of research and heated discussions on Facebook and Instagram, atmosfair (www.atmosfair.de), who take into account the impacts of non-CO2 aircraft emissions and Mossy Earth (www.mossy.earth), who focus on native tree planting and rewilding, are the best options. Though offsetting will cost you, it seems that if you can afford to fly, you can afford to offset it. Now, let's get back to saving the world for free and discover a few simple ways to make unavoidable air travel a bit greener.

FLY NON-STOP AND ECONOMY

Choose a non-stop flight if you can, as take-offs and landings are really 'guzzly' when it comes to fuel. And it's far better to fly economy, as first-class seats take up around twice as much space as an economy seat. Which means someone travelling first class is responsible for twice as much carbon as someone flying economy! Choosing budget airlines takes this even further, as with no first- or business-class option, they have smaller per passenger carbon footprints.[14]

If you can, take the slow road and go by bus, especially if you are travelling between cities in the same country. An eye-opening and easy-to-read report by the Union of Concerned Scientists (gotta love that name), showed that coach travel is by far the greenest way to travel, even for a family of four. And it'll cost you less than flying.

TOP-UP TRAVEL TIPS

Packing to go on holiday these days often involves a certain amount of messy-but-satisfying decanting of shampoos, conditioners and shower gels into 100ml sized bottles to get you through airport security. If you're not already refilling your own travel containers, then make this one of the small steps you take to help save the world. According to a study commissioned in the UK in 2018, British holidaymakers left 980 tonnes of travel-sized containers behind them at their holiday destinations the previous year, and paid up to eight times more for them than for full-sized versions.[15] Many of the countries we're visiting have even worse recycling facilities than we do; for example, the arrival of 200 million tourists in the Mediterranean each year causes an alarming 40 per cent rise in plastic waste in the sea every summer.[16]

So, refill your own travel bottles and take them back home with you for your next trip, or better still, go packaging-free and take solid versions of soap, shampoo, sunscreen, deodorant and toothpaste with you instead. (See 'Save the World When You...Use the Bathroom' on p.131 for more on toiletries.) The airport security staff won't believe you when you say you have 'no liquids', but think how smug you'll feel when you show them your range of solid products, with not a leaky plastic lid in sight.

And don't forget your water bottle. Carry an empty one with you or empty it out before you go through customs and get a free refill from a fountain, shop or café once you're through. Pledge not to buy water in single-use plastic while you're away and, instead, find refill points (you could even add them to the Refill app!) while you travel. Check to see if the tap water is safe to drink before you go and, if it's not, get yourself a filtration bottle to filter water for you on the go. You'll still save yourself a packet by not buying bottled water and you'll be easing the load on the local environment. And reusable bottles are a great talking point – you may well find yourself extolling their virtues to fellow travellers and locals.

BETTER BACKPACKING

Whilst staying at an eco-lodge may not qualify for 'saving the world for free', if you're planning a holiday why not make it an active one, and sign up for a volunteering scheme abroad? Lots of destinations offer free accommodation in return for a few hours' work each day and you meet people, get fit and get to know the local environment in ways you'd never experience if you were 'just a tourist'. From helping to build a school to reforesting tropical rainforests (probably the best thing you can do on holiday to combat climate change), or taking part in ocean-conservation projects, you can get your travel fix while simultaneously making the world a better place – and maybe even learn a new language while you're there. Eco-tourism schemes help protect biodiversity and allow countries and communities around the world to build resilient economies without harming the environment.

And if you find yourself saying, 'But I need a holiday, not more work!' remember that helping others makes you happier, lowers blood pressure, helps you live longer and creates ripples of goodness – one study found that people are more likely to 'perform feats of generosity' after observing another do the same.[17] This type of travel is also a fabulous way for school-leavers and teenagers to see the world.

MORE WAYS TO SAVE THE WORLD WHEN YOU TRAVEL

CHOOSE A STEEL GRANNY BIKE, OR SECOND-HAND RETRO ROAD BIKE IF YOU WANT EXTRA GREEN KUDOS: steel only releases a third of the carbon-dioxide emissions of aluminium in the production process and is 100 per cent recyclable.

ECO-FRIENDLY DRIVING: if you have to drive, there are lots of little things you can do to reduce fuel use. These include keeping tyres properly inflated to maximize efficiency, driving and accelerating slowly (going over 95kmh [60mph] greatly increases fuel use), opening the windows rather than using the air con, switching off the engine during waits, driving a smaller car, using a sat nav so as to take the shortest route, keeping air filters replaced to improve mileage, and using cruise control.

AVOID RUSH HOUR: stagger your work hours to avoid times of peak traffic, so you won't be sitting idling your engine in jams, wasting fuel and increasing emissions, as well as stress levels!

WALK: for shorter journeys, brisk walking will provide you with your daily aerobic exercise quota and costs nothing. If you are going food shopping, carrying the bags home rather than putting them in the car will provide an extra workout and help you tone your arm muscles.

ENJOY A STAYCATION: a family of four flying abroad for a holiday will create several tonnes of carbon dioxide (for example, from the UK to Disney World in Florida generates 9 tonnes[18]), so why not explore the beauty spots in your own country?

GO WWOOFING: World Wide Opportunities on Organic Farms is a global movement that enables volunteers to work on environmentally conscious organic farms, gardens or woodland settlements across the globe in return for free bed and board.

TRY AN ECO-HOLIDAY: there are many organizations offering expeditions to plant trees and do conservation work. Getting there is going to produce carbon, but if you are after an experience in another country, this will offset some of your footprint. Look for seals of approval from reputable certification programmes or affiliations with international organizations such as

WWF (World Wide Fund for Nature), Rainforest Alliance or Nature Conservancy, or, in the UK, the Green Tourism Business Scheme.

AVOID CRUISES: cruise liners consume the same amount of fuel as small towns – the biggest can emit more sulphur than several million cars – and, collectively, they spew out 1 billion tonnes of sewage each year into the pristine environments that travellers (25 million of them annually) have gone there to enjoy.[19] Enough said.

WHILE YOU'RE AWAY: if renting a car, choose a hybrid or electric if available and keep it as small as possible, keep your showers short and don't let hotels wash your towels and sheets every day – once a week is enough.

CAMPAIGN FOR A FREQUENT FLYER TAX: Despite being the fastest growing cause of climate breakdown, there's no tax on jet fuel. In fact, it's the only fossil fuel that's banned from being taxed by international treaty. This needs to change, fast. Find out more and join the campaign to replace the current tax on flights with a fairer system that taxes people according to how often they fly at afreeride.org.

Save the World Where You...Live

The idea of a 'green home' may well conjure images for you of solar panels, fancy Scandi-style architecture and underfloor heating. Which, admittedly, sounds expensive. Yet, in fact, some of the things you can do to 'green' your home can make a huge difference without you having to take out a second mortgage, get a bank loan or call the builders in. Actually, a whole load of them won't cost you anything.

WHEN GREENER IS CHEAPER

Heating homes (or cooling them, if you live in a hot place) usually involves fossil fuels. 'But switching to a green-energy supplier is more expensive!' I hear you cry, before I've even written the words, so we should all switch to a green-energy supplier. Well, let's address that right away, as things are changing.

Almost half of consumers in the UK who have switched to a green tariff have done so to save money, according to a survey conducted by moneysupermarket.com. Some specialist energy suppliers in the UK are beginning to offer more competitive prices and, thanks to the trend towards renewable energy, the costs should continue to fall. Hurrah! It's not uncommon now to find that the cheapest switchable tariff on moneysupermarket.com has been to 100 per cent renewable energy. So, get switching, get saving and get yourself, your business, your school, and whatever else you can manage, onto a green energy tariff. Then, of course, there's that obvious solution to saving energy (and lots of money) at home: just use less of it. Below are just some of the ways you can cut down on energy consumption.

WASHING CLOTHES

Assuming you have a washing machine, washing your clothes at 30 instead of 40 degrees Celsius can be one third cheaper, saving up to £52 a year (or changing from a warm to a cool cycle and saving up to $70 in the US)! And if, for example, every Londoner switched to 30°C, they would save enough energy between them to rotate that giant Ferris wheel, the London Eye, 2.3 million times.[1] You could go even further and – if you're not washing nappies or reusable menstrual pads – experiment with a 20°C wash (or a cold one, in the US). It works for me – I've been doing it for a year and haven't noticed the difference; the clothes still come out nice and clean. And, at 20°C, most colours (at least from clothes that aren't brand new) won't leach, so you can avoid wasting power on half-full loads. If you're always sticking a small load on because you keep running out of undies then maybe it's time to invest in some more – organic bamboo ones, preferably. You can afford them now you're saving on your electricity bills! Oh, and try using soap nuts (dried fruits of a tree that grows in India and Nepal which contains saponins that clean your clothes when mixed with water) instead of detergent for washing clothes – you can get a hundred washes from one small bag.

SWITCH OFF AT NIGHT

Switch off your appliances (but obviously not your fridge or freezer) at night, at the socket – it can save you up to £80 a year (around US$40–105), depending on how many gadgets you have.[2] It's one of the simplest actions you can take in this entire book. Here's how to have some fun by switching off:

HAVE AN ELECTRONIC SUNDOWN

At a certain time each night (an hour before your usual bedtime is good) go around your house switching all the sockets off. You could try putting stickers by your sockets with life-affirming sayings written on them, to give you a warm inner glow. For example, the famous hypnotherapist, Marisa Peer, has written a brilliant book on the power of the words, 'I am enough', so you could try sticking those three words above the plug sockets for a month and see how you feel. It might even make you sleep better.

BEE KIND, TURN YOUR WIFI OFF AT NIGHT

Studies have shown that people sleep better when the WiFi is switched off[3] and given that the World Health Organization has classified WiFi in the 2B group, alongside lead, as a possible human carcinogen, then we're all probably better off erring on the side of caution since no studies have, as yet, been done on long-term 24/7 exposure. The bees will probably be grateful too: electromagnetic radiation has been found to alter bee behaviour, produce biochemical changes and impact bee reproduction. Studies on the effect of electromagnetic fields from cell towers and wireless devices concluded that out of the 919 research papers collected on birds, bees, plants and other animals (including humans!), over half showed negative impacts.[4] If this makes you want to switch it off in the day too, a wired dLAN (direct Local Area Network) connection is easy to set up and will free up your airwaves from radio-frequency radiation. Sweet dreams all round.

HAVE A GREAT BIG GREEN RE-FIT

If you're canny, or have an abundance of cash and own your own home, you can do a major retrofit of your house. Solar panels and tiles, Tesla batteries, biomass boilers, geothermal heaters and sedum rooftops are just a few of the wonderful ways you can save energy.

LOW-CARBON HOUSING INITIATIVES

There are some exciting low-carbon housing initiatives popping up around the world, such as the UK-based Citu, as well as community-led housing schemes. If you're about to hop on the housing ladder or climb another rung, have a look online for community land trusts, housing co-operatives, self-build schemes and cohousing in your local area to see what future-loving, planet-friendly alternatives are on offer where you live.

RECYCLE, FREECYCLE

Next time you break something, or move house, instead of racing to your nearest furniture store, spending more than you need and breaking into a sweat (I call this 'Ikea syndrome'), try your luck online for something preloved to help protect the earth's resources. Ecologist Trevon Fuller links the 'unbelievably high demand' in the US for furniture made in China to deforestation in Africa's Congo Basin, with the export of wood from the Congo Basin to China doubling between 2001 and 2015 according to his 2018 study.[5] Here's an example of how our actions can make a direct difference: quite simply, buying second-hand furniture instead of new Chinese furniture would save you money while keeping more of the Congo Basin's rainforests intact. Who wants to eat supper around a table that used to be home to a family of mountain gorillas anyway?

A quick search for your local equivalent of Freecycle, Gumtree or Craigslist could result in you finding the perfect thing. It will more than likely be cooler and cheaper, and it will definitely be greener. You'll also get to nose around someone else's house and maybe even make a new friend.

SOAK IT UP WITH GREEN SPACES

Planting more trees and flowers, and creating and protecting our green spaces, is good for the planet. We all know that. One of the ways green spaces can do this is by reducing urban runoff. That's not, as it might sound, a brand of sportswear, but one of the biggest threats to water quality in urban areas. Without green spaces to soak it up, rainwater has nowhere to go, so it pours off rooftops and roads into storm drains and, from there, into our rivers and seas, picking up masses of polluting chemicals on the way. We add to urban flooding when we do away with our sponge-like soil. Gardens are gorgeous, so have a look at your home and office and see if you can introduce more greenery to help reduce runoff. Even a few outdoor potted plants can help (you can often get the pots free on Freecycle or your local household recycling centre – plant them with bee-friendly plants for maximum planet-saving effect).

If the drainpipes from your roof are connected directly to a storm drain, disconnecting them is the single most important step you can take to reduce runoff.[6] So hurry up and divert the water running off your roof into water butts to water your budding vegetable garden (because you're starting to grow your own, right?), or just redirect the pipes towards your garden or lawn. Which leads me nicely on to the next section.

GREYWATER IS GREEN

Aside from helping to prevent ocean pollution, water shortages and flooding, harnessing the rainwater from your roof and the greywater from your sinks, baths, showers, dishwashers and washing machines is a *must* for any world-saving wannabe. Greywater (not the brown or yellow water that comes from your toilet) makes up between 50–80 per cent of a household's waste water[7] and really doesn't need to travel miles and miles to go through an expensive treatment process when it could, literally, work wonders in your garden or on your houseplants. But unless you're building a new eco-home it can get a bit complicated and expensive to re-plumb your house or apartment, so a simple solution is to throw your washing-up water or bathwater in the garden. Try to stick to natural cleaning products, since you don't want a load of chemicals in your soil. Water shortages are increasingly common and, as the planet heats up, you're going to need to make every drop of water count, especially if you're growing some mighty-fine vegetables in your garden. Using your greywater saves you money too, especially if you're on a metered water supply!

BREATHE EASY AT HOME

Indoor air pollutants have been ranked in the top five environmental risks to public health.[8] And we can't see them! It's a big enough problem inside our homes and schools (and our spaceships) that NASA got onto it and found a simple solution: houseplants. Indoor plants, along with the microorganisms living in their soil, clean the air of impurities and chemicals found in building materials, paint and cleaning agents, such as formaldehyde and benzene, so act as electricity-free air-filtration systems. They also convert carbon dioxide into oxygen, helping us to breathe easy. Houseplants are lovely to look at, too, and create a sense of wellbeing, with one study showing that hospital patients with plants in their rooms reported lower stress levels and had lower blood pressure.[9] Plants that have been found to be especially efficient at neutralizing nasties include chrysanthemums, dracaenas, snake plants, peace lilies, aloe vera, spider plants, English ivy, Boston ferns, orchids and weeping figs and rubber plants.

CLEAN BUT GREEN

Next time you get out one of the (plastic) bottles from that collection under the sink, you might want to ask yourself if you really want to buy another one when it runs out. Household cleaning chemicals all end up in our soil, water and even our air in the end – a recent look at the air in Los Angeles showed that volatile compounds from things like perfume and surface cleaners were as much a problem as vehicle emissions.[10] Weedkiller, too, ends up in our soil and waterways, with around 40 million kilograms (90 million lbs) of it being applied every year to lawns and gardens in the UK alone. Many of the chemicals in weedkiller have been classified as probable carcinogens.[11] The crazy thing is that we really don't need most, or even any, of these products – elbow grease and water can work just as well. Vinegar, lemon and baking soda all do a great job of removing grime and can all be bought plastic-free, whilst a sharp trowel can deal with unwanted weeds and do away with the need for weedkiller.

GROW YOUR OWN

A book about saving the world for free wouldn't be worth its weight in recycled paper if it didn't make you feel very, very inspired to start growing your own food. So, watch me try. If I fail to ignite your inner gardener's spark, try reading this chapter in a couple of years' time. You might not be ready for it just yet. And that's ok. You may have travelling to do, sofas to surf on and adventures to go on before you find yourself with a windowsill to call your own.

Whether it's your own coriander, chillies and tomatoes grown on the aforementioned windowsill (did somebody say 'salsa'?) or a supply of squash that could feed your entire neighbourhood, growing your own food is a wondrous way to save the world, for free. How so? Well, by the time you've begged, borrowed and re-homed everything you need to get set up (as well as pinching the odd sprig of rosemary from a bush overhanging the pavement) you can garden for no cost. Seed swaps are a great way to maximize the range of veg you can grow by sharing your leftover or unwanted seeds with other local gardeners. And, once your garden is producing, you can always swap and share your harvest too. Got too many French beans? Pickle them or swap them with your neighbours for whatever they've got too much of. Got a glut of courgettes? Cook them up in a curry, portion it out and freeze it for enjoying on a cold winter's night.

Gardening is good for you too. It relaxes us, gets us outside and gives us something to nurture. When I started gardening, admittedly not until later in life, I thought I would burst with happiness. As someone who thinks a lot, probably too much, getting my hands dirty, being physically active and being outside – away from screens and social media – made me feel good, peaceful and happy-tired. I've had a few epic fails; not having a clue what I was doing meant that some things didn't grow, or they were ravished by slugs or snails. But the first year I grew kale successfully and had fresh, organic greens throughout the whole winter was a triumph. And that was in a tiny garden too.

You don't have to devote your life to growing your own, unless you want to, of course. If you have a small patch of garden to cultivate, then an hour or two at weekends, plus some casual watering and tending during the week will be ample.

MORE WAYS TO SAVE THE WORLD WHERE YOU LIVE

CALCULATE YOUR PERSONAL FOOTPRINT: see if you're using more resources than the planet can provide and use it as a guide for cutting down. Try the Earth Overshoot Day calculator at footprintcalculator.org.

CHOOSE SUSTAINABLE FURNITURE: seek out items made from bamboo, recycled plastic and wood, or that have a sustainability certification process.

USE FISH WATER FOR PLANTS: if you have a fish tank, reuse the dirty water on your houseplants. Plants will love it!

BREAK DISHWASHING TABLETS IN HALF: your plates and cutlery will come out just as clean. And load your dishwasher consciously: don't, as some people do, put three pans and a wine glass in it and then set it off.

USE LED LIGHT BULBS: they can be up to 80 per cent more efficient than other bulbs and last up to six times longer.[12] They aren't perfect – despite being classified as safe for landfill, they contain toxic elements including lead, but they are, at least, preferable to mercury-high incandescent bulbs.

PUT SOME THERMALS ON: warm up with layers or a sweater instead of turning the heating up.

DRY YOUR CLOTHES NATURALLY: use a washing line, drying rack or clotheshorse instead of a tumble dryer.

AVOID FRIDGE-GAZING: there won't magically be something there that wasn't there the last time you looked, so there's little to be gained from standing gazing hopefully inside the fridge with the door open. And, keep your refrigerator in the coolest part of your kitchen – it will use less energy that way.

TURN OFF THE TAP: save water while brushing teeth by not turning the tap on until you rinse. And fix any leaky taps. A leaky tap can lose 20,000 litres (5,300 US gallons) of water in a year![13]

BEE-FRIENDLY IN YOUR GARDEN: if you have any outdoor space, make sure to plant it up with bee-friendly plants such as lavender, cotoneaster, heather, chives, catmint and foxgloves.

REFILL EVERYTHING YOU CAN: laundry liquids and cleaning products can often be refilled from a health-food shop, making them cheaper than new ones. Or search online for refillable, eco-friendly detergent deliveries.

DONATE LEFTOVER PAINT: unused paint can be given to a local community scheme instead of adding to the millions of litres of paint thrown out every year.

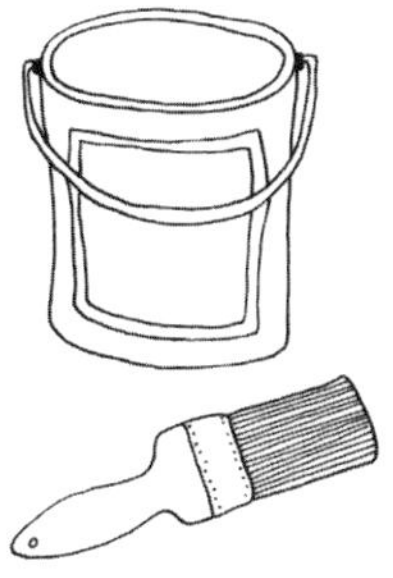

BE LIKE THE DANISH, DON'T DO DRAUGHTS: you can help keep your home warm by insulating it with double-glazing, by draught-proofing and by insulating lofts, cavity walls, floors and behind radiators. Danes are so good at this that they don't even have a word for 'draught'.

CANCEL JUNK MAIL: help stop deforestation by cancelling mail you no longer need and stick a 'No leaflets, menus or junk mail' sign on your door.

GIVE AWAY YOUR STUFF: check out www.trashnothing.com when you need to downsize, move house or are having a declutter. It's a combo of Freecycle, Freegle and others!

GET BURIED IN A GREEN COFFIN: these are made from a material such as willow or bamboo. Or have your ashes turned into a living, breathing, carbon-absorbing tree with a bio urn.

Save the World When You...Bank

Wealth inequality, boom and bust financial markets and longer working hours may not seem relevant to a book on green living, but this is a book about saving the world. Understanding a little more about how our world operates helps us to get some context for the ecological mess we're in, especially when it comes to our 'need more stuff' consumerist culture. And, while I'm no finance expert, it's plain to see that economic growth is directly linked to increased pollution and consumption of the Earth's resources. This section is unlikely to reform the monetary system, but it will give you some pointers on how that *could* happen and might just get you questioning who your bank is really serving, as well as giving you some simple ways to influence what your bank or your pension fund does with your dosh.

MONEY HONEY

Banks use the money you're lending them to lend out to others, for profit. But do a little digging into how they make that profit (and how they spend it), and you'll be left quivering in your wellington boots. Most of the investments made by the 'big banks' are far from ethical. Put simply, if you're with any of the big banks, your money is funding war (through the arms trade), climate change (through industrial cattle farming and oil and gas extraction), and ocean pollution (through harmful pesticides). And, if all that wasn't enough to make you run as fast as you can towards the (UK) government's Current Account Switch Service (CASS), the rise in global income inequality can also be blamed on the soaring pay of CEOs and Wall Street Bankers, according to economists at the Economic Policy Institute.[1]

ETHICAL BANKS – A PENNY-DROPPING MOMENT

Switching to an ethical bank, and discovering the environmental benefits of doing so, was a massive lightbulb moment for me. I've never had large enough amounts of money to think that where I kept it, or moved it, was of any consequence. But how wrong I was! And how gleefully I switched when I realized that, by using an ethical bank (of which there are very few), I would be helping to stop plastic pollution, fracking and other pet environmental peeves of mine, at source.

DOING GOOD WITH YOUR GREEN

Banking ethically is a simple, yet powerful, way to vote against environmental destruction and oppression. The Dakota Access pipeline, near the Standing Rock Sioux reservation, was a potent example of this, as the crisis galvanized people to withdraw their money and switch away from the banks funding the pipeline. Thousands of people responded and DNB, Norway's largest bank – one of the main investors in the project – pulled its assets.[2] We are that powerful. Who you bank with matters. Don't forget to send a letter, email or message on social media to your bank when you switch to an ethical one. Make them aware that you're leaving their poor practices and policies behind and telling all your friends as you go.

ALL THAT GLITTERS

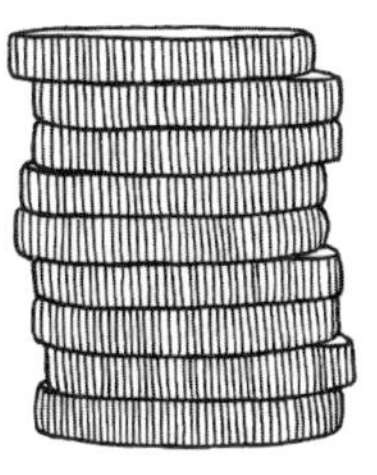

It's worth saying at this point that the bright new shiny brands in 'fintech' (financial technology) such as Monzo, and some new banks on the scene such as Atom and Starling are appealing because of their design and ease of use, but given *you* want to save the world for free you might want to do some (gold-) digging before you jump in. As profit-driven (not mission-driven) businesses, often backed by venture capitalists, there is nothing forcing them to behave ethically other than a fear of getting caught by regulators. So a savvy question to ask when greening your banking would be 'what is baked into this bank's ownership, governance, or mission that means it will always put people and planet before profits?'

PLANET-FRIENDLY PENSIONS

Let's not forget to mention pensions. If you're an employee, you're most likely part of a pension scheme. Happily, a growing number of pension funds across the world are now divesting from companies that generate revenues from oil, gas and coal. To check yours, log in to your pension online and search for 'fund choices'. It could take you just one incredibly well-spent minute to make the switch to an ethical fund within your company pension scheme.

If you have, or are looking for a personal pension, look for terms such as ESG (which stands for environmental, social and governance) or SRI (socially responsible investment) as these forms of investment actively look for companies that are part of creating a better future. Which makes sense, as saving for a future by giving money to things that threaten it makes no sense whatsoever.

HOMEGROWN CURRENCIES

As well as choosing an ethical bank, there could also be a homegrown currency on your doorstep. There are currently around 300 of these schemes around the globe, some of which are shining examples of how a local currency can create more resilient communities. Local currency won't cost you more than your conventional currency, and may even save you money, as lots of independent shops on local-currency schemes offer discounts. And the notes often look amaaazing too.

Local currencies create wealthier, more stable, greener communities because the money stays in the community,

supporting small local traders instead of leaking out into complicated global financial systems. By supporting local businesses in sourcing more goods and services locally, using a local currency also reduces their carbon footprint. In the UK, the Brixton Pound, in South London, supports 250 local businesses which could otherwise be under threat from global chains. The aim of the Brixton Pound is to stop 80p of every £1 spent locally from leaving the area. Meanwhile the Bristol Pound (in my home city), created in 2009, now links a robust network of over 2000 individuals and independent businesses. In the US, BerkShares are the popular local currency for the Berkshire region of Massachusetts. Launched in 2006, over 7 million BerkShares have been issued and accepted by more than 400 businesses in Berkshire County.

AN ALTERNATIVE TO CAPITALISM

You've heard of the phrase 'the rich get richer while the poor get poorer'? Well, that's still the case. The world's richest 1 per cent own almost half the world's wealth[3] and, according to Oxfam, eight men own as much wealth as the 3.6 billion people who make up the poorest in the world – while one in ten people survive on less than £1.50 (US$2) per day. Thanks to capitalism, global wealth inequality is still on the rise and businesses rule the world. Of the 100 largest economies in the world, 69 are corporations.[4] And, just to rub a sprinkling of salt in that wound, it's worth noting that there may be as many as four times more psychopaths amongst CEOs than there are in the population at large.[5] So, potentially a lot of our corporations are led by people who lack vital emotions: empathy, remorse and loving kindness. *Now* do you see what all this has to do with restoring and protecting this precious planet we call home?

As it stands, producers and consumers don't have to pay for the costs of the pollution that comes from their decisions. The drive by these corporations for short-term profits comes at a cost to basic human rights as well as the rights of nature. (Which, thankfully, some countries like New Zealand are starting to acknowledge – in 2017 the Whanganui river was granted the same legal rights as a human being.)

ECONOMICS FOR YOU, ME AND THE TREES

Participatory economics is an alternative model, being explored and seeded around the world, that gives everyone the opportunity to influence decisions about themselves, in proportion to how much they're affected. Local citizens belong to neighbourhood councils and get to vote on decisions about consumption and local public goods. Unlike our hierarchical systems, participatory economics is built on democracy and nature conservation – it understands that continued economic growth on a planet that's not growing just doesn't work. Especially not for the poor. A system that supports individuals and communities getting as much power in their own lives, and as little power over others' lives, as possible, is an exciting alternative to capitalism. Power is given back to people so that wealth isn't continually given back to the rich. If you can't find a free workshop or talks on participatory economics in your local city or town there are loads online to get you started. Or have a read of the chapter 'Save the World When You...Vote' on p.181 to see how participatory budgeting systems are already being used around the world.

A SWEETER SYSTEM

Another inspiring roadmap for an economic model that works within our environmental limits is the concept of Doughnut Economics, created by Oxford academic Kate Raworth. If systems are your thing, and you love the idea of understanding and expanding ways to help humanity thrive regardless of whether the economy is growing or not, then I recommend stuffing your brain with *Doughnut Economics*. Grab a copy of her book from your local library, get inspired, and save the world from our broken economic system, for free.

VOTING WITH YOUR WALLET

Just one more thing on capitalism: there's a very easy way to break the system and that's to be happy. Our economy only works because we keep buying so much stuff. Corporations spend billions on advertising and marketing to make us feel rubbish about ourselves, so we buy their products to feel better. Being happy is a revolutionary act; happiness cuts off capitalism's balls and uses them as compost to create a better future.

MORE WAYS TO SAVE THE WORLD WHEN YOU BANK

GET CURRENT AND SWITCH YOUR SAVINGS: Triodos (currently only operating in parts of Europe) is the best example I know of a genuinely ethical bank that has a current account as well as saving and investment options. Every organization they finance is making a positive difference – socially, culturally or environmentally. And they're transparent, meaning they publish details of who they lend to and invest in.

INVEST IN THE FUTURE: If you're looking to invest and want to do good with your money, check out sustainable investment and crowdfunding platforms such as Ethex, WorthWild and Thrive Renewables for some great opportunities.

OWN YOUR OWN BANK: Another alternative to profit-driven banks are credit unions, which have a positive social and environmental impact by supporting their local community.

GO FOR GREEN BONDS: US readers can switch traditional bonds to green municipal bonds; proceeds are used to fund environmentally friendly projects and their popularity is on the rise. In 2017, $155 billion were issued compared to over $250 billion in 2018.[6]

STAY LOCAL AND COOPERATE: Look into what community banks are operating in your area: these are locally owned and operated and focus on the needs of local businesses, families and farmers.

UNDERSTAND THE SYSTEM: Check out positivemoney.org to understand more about our financial systems and find out ways to become more involved in your community.

SHARE YOUR LEARNINGS: Arrange a film night with friends to watch the documentary films *Four Horsemen* and *97% Owned*, then have a conversation about how our financial systems affect our environment.

MAKE A STATEMENT, GO ELECTRONIC: Save paper by switching to paperless bank statements.

Save the World When You...Dress

Regardless of whether or not fashion is important to you, we all wear clothes, so we need to get savvy about the human and environmental cost of this trillion-pound industry – the lives that are being lost, the land that's being devastated and the rivers that are being contaminated. The clothing business is a very dirty one – it's the second biggest polluter of our world's water after agriculture, due to the pesticides used in growing cotton and the vast number of toxic chemicals used in production.[1] It's also responsible for 10 per cent of global carbon emissions, a figure projected to increase by more than 60 per cent to over 2.5 billion tonnes per year by 2030.[2] We need to slow fast-fashion down.

Use this as a reason to invest in some high-quality, organic, locally made or Fair Trade clothes. Even if they cost more to begin with, they'll save you money as they'll last longer and they'll be a shining statement of your commitment to looking and feeling great, while supporting conscious, ethical fashion brands.

SWEET CHARITY?

'Charity shops', 'thrift stores', 'op shops' or 'chazzas'. Call them what you like, second-hand is chic if you want to save the world. Charity shops may not be free, but they'll save you a lot of money compared to buying new, and they'll save a lot more than just money: using charity shops is an easy way to

reuse and recycle clothes; they help unemployed people gain skills and work experience through volunteering; they create jobs, they bring life to high streets by using empty space; and they raise a lot of money for good causes. In just the UK, £270 million pounds (about US$354 million) in profit was generated for the period 2015–16.[3] And – here's the green part – almost 7 million tonnes of carbon-dioxide emissions were avoided by charity shops in that same period simply by diverting clothes from landfill.[4] That's the equivalent of two years' worth of emissions from the entire country of Iceland.

But let's not think that charity shopping holds all the answers. In the US, only 10 per cent of clothes that are donated to charity shops are resold, and in Australia the figure is around a third. The rest either go to landfill or are exported to developing countries – according to Oxfam, more than 70 per cent of the clothes donated globally end up in Africa, which greatly undermines their own garment-manufacturing industries.[5] So, if you find yourself shopping for new clothes on a weekly basis, and regularly taking your unwanted items to charity shops, you're still part of the problem.

COTTON ON

It's unlikely you'll be able to source your entire wardrobe second-hand, especially when it comes to your underwear. (You probably could source it second-hand, but you may not want to.) So, a nice new pair of natural cotton knickers or boxers would be the obvious choice, right? Well, not necessarily.

OUR DIRTY COTTON CLOTHES

Cotton is officially the world's 'dirtiest' agricultural commodity, because it uses a staggering 16 per cent of the world's insecticides, which not only kill insects but also cause widespread health issues and birth defects in the communities using them.[6] Another problem with cotton is that it is grown in warmer countries, where water is already scarce and getting scarcer. The water used to grow India's cotton exports in 2013 would have been enough to supply 85 per cent of the country's 1.24 billion people with 100 litres (26.5 US gallons) of water every day for a year, according to ethicalconsumer.org. That year, more than 100 million people in India had no access to safe water. In addition, over 95 per cent of cotton grown in the US and India is genetically modified, which causes problems with pesticide resistance.[7] GM cotton has also been linked to farmer suicides. This is thought to be due to pressure resulting from companies having monopolies over the GM seed, forcing farmers to buy new seeds every year instead of being able to reuse seeds saved from the previous harvest.[8]

NYLON'S NOT THE ANSWER

Nylon and polyester are not the answer, unfortunately. They're made from fossil fuels – a carbon-intensive, non-renewable, non-biodegradable resource. And nylon production releases nitrous oxide, a greenhouse gas that's 300 times more damaging than carbon dioxide. Then there are the harmful chemicals used to produce polyester, including dangerous carcinogens which are often discharged into local water supplies by manufacturing plants in countries such as China and Bangladesh (where environmental regulations are less stringent), resulting in significant pollution downstream. Those same chemicals – nasties such as dimethylformamide, azo dyes and phthalates – are also in the clothes and can be absorbed into our skin.

MICROFIBRES

And the pollution doesn't end with synthetic clothing. It's coming out of our homes too. It's estimated that a single piece of clothing made from synthetic material, such as polyester or nylon, releases up to 1,900 microfibres each time it's washed.[9] Microfibres are tiny fragments of thread that are shed from our clothing when we wash them. They're the third biggest contributor to microplastics in the ocean and, being so small, they pass through sewage filters and wastewater treatment processes. One study showed that, from 18 beaches worldwide, 85 per cent of the human-made material found on the shoreline were microfibres that matched the types of material, such as nylon and acrylic, used in clothing.[10]

In the ocean, microfibres act like a sponge for all toxic chemicals, turning them into dangerously poisonous morsels which little fishies just love to eat – they seem to be like junk food is to teenagers. The added problem here is that not only are microplastics being consumed by plankton and tiddly fish, they're making their way up the food chain and into the fish and shellfish we humans eat. While studies have yet to show us the effects of this on human health, they do show that toxins leaching from plastics into fish can move all the way up the food chain into our bodies. If you're an oyster or mussel lover, one study estimates you're eating up to 11,000 pieces of microplastic a year.[11]

Enough of the bad news? I thought so. Let's find some solutions and get shouting them from the rooftops.

OPT FOR ORGANIC COTTON

Instead of buying lots of clothes that screw the planet over, buy fewer and spend the money you've saved on some quality, organic additions to your wardrobe. Organic cotton, while still water-intensive, is grown without all the nasty pesticides and makes for a safer growing environment for the workers and bees alike. There are lots of organic cotton clothing producers online.

BUY BAMBOO

Bamboo is an eco-friendly alternative to cotton and synthetic fibres. It improves the soil quality instead of depleting it, it's fast-growing, it doesn't need to be replanted and it can be grown without fertilizers or pesticides. Yay. But look out for 'bamboo-based rayon' as this uses toxic chemicals to turn the plant into fabric. Choose organic bamboo threads to stay on the right side of happy. Especially if you're in a relationship – my partner says that I stroke him more often when he's wearing a bamboo shirt. (It feels so soft though!)

SAY HELLO TO HEMP, LOVE LINEN AND TRY TENCEL

Hemp is another of the plants that can save the world. It's drought-resistant, it can be grown in most climates, it won't deplete soil nutrients and it's easy to harvest. Linen, made from flax, uses less water in production than cotton and can be grown without pesticides. And Tencel, made from sustainably sourced wood, is another natural fibre that is winning awards for its closed-loop process that recovers or decomposes all solvents and emissions. Ok, now we're talking.

FAIR TRADE ALL THE WAY

Finally, I'll give the last word here to Fair Trade. It may not always be organic, but you can bet your bottom dollar (pound or euro!), that your money is ensuring better pay and working conditions for the people making your clothes, while helping protect the planet – and that goes for your coffee as well as your clothes.

BE ORIGINAL – LOVE PRELOVED

Keep an eye out for 'preloved' and 'dress agency' shops for second-hand designer and high-end clothing when you want to turn clothes you've grown out of or grown tired of into cash, or need something a bit special and don't have time to rummage around your local charity shops. I adore my local one. As well as reducing carbon-dioxide emissions by reusing and recycling, shopping second-hand is a statement of who you are – you buy clothes that you love and look good on you regardless of what's in fashion this season (or this week).

A lot of us keep shopping for new stuff because it feels good. We get a surge of dopamine when we hit the shops, which is our brain's way of anticipating a reward, and it also spikes when we see something reduced or on sale. Hence the frenzied feel of outlet stores and shops such as TJ/TK Maxx, which, by the way, are a total con – most of the clothes you find in outlet stores were never actually on sale in their designer stores, they're just cheap versions created specifically in partnership with outlet stores to trigger our 'shopper's high' and trick us into thinking we've scored a bargain.[12] My TK Maxx bubble has well and truly burst.

Feeling good is important though, so below are a few feelgood things you could do instead of lining the pockets of planet-polluting clothing brands. When you next notice the urge to splurge try one of these.

ORGANIZE A CLOTHES SWAP

According to resource-efficiency experts WRAP, Brits have an estimated £30 billion (that's US$46.7 billion) worth of unworn clothes lingering in their wardrobes.[13] Let's just ponder that for a moment. £30 billion. Divided by the population of the UK that's £450 each (roughly US$590). Of unworn clothes?! Now a clothes swap is sounding like an even better idea. You'll make friends, save money and make a difference. Oh, and probably still get your dopamine fix as you rummage gleefully through a pile of free clothes. Hooaaaah!

DECLUTTER AND RE-ORGANIZE YOUR WARDROBE

And, when you've decluttered, treat yourself to some cake! A 2018 study with 18,000 heads of households in 20 countries, showed that Americans had the second highest levels of unworn clothes in the world, wearing less than a fifth of their wardrobe.[14] In case you're wondering, Belgians are the worst. And Brits aren't far behind. So, have a clear out – invite a friend round to help you be decisive, to make it more fun and to give you a second opinion. And give a couple of bags of your old threads away to your local charity shop. Decluttering can give you more headspace and a fresher-than-fresh feeling every time you open your wardrobe. Another great tip is to turn all the hangers the same way around, then, after you've worn an item, return it to the wardrobe with the hanger pointing the opposite way. After a year, you'll be able to easily spot anything you've not worn. The cake bit is just a reward really, although if you've baked it yourself (with all the time and money you've saved yourself by not shopping), and you give some to your neighbour or take some to work, your happiness levels will shoot through the roof.

MORE WAYS TO SAVE THE WORLD WHEN YOU DRESS

DOWNLOAD THE 'GOOD ON YOU' APP: it'll give you some instant insight into how well over 2000 fashion brands are doing when it comes to looking after people and the planet.

ORGANIZE A HOME-SCREENING OF *THE TRUE COST*: the 2015 documentary shows us the truth behind the fast-fashion business and how it affects the people making our clothes.

SUPPORT CHARITIES THAT SUPPORT WORKERS' RIGHTS: following and sharing content from organizations such as Labour Behind the Label is a free and easy way to raise awareness.

USE REPAIR SERVICES: some pioneering companies offer a repair service to make their clothes last longer. Patagonia, who make sustainable outdoor clothing, do this and also offer DIY repair guides for their clothes.

WEAR IT 50 TIMES: wear each item of clothing you own 50 times instead of the average 5 times to minimize carbon emissions from buying clothes you don't need.

UNFOLLOW CELEBRITIES THAT PROMOTE FAST-FASHION: follow people who are changing the world for the better instead.

MINIMIZE MICROFIBRES WITH A MESH BAG: Guppy Friend are leading the way in mesh bags designed to hold your clothes inside the washing machine to catch microfibres from synthetic clothes. You can order these online, but some brands have also started selling them with their clothes.

BUY SUSTAINABLE FASHION: look out for certification, such as the Global Organic Textile Standard (GOTS) or Oeko-Tex, which certify socially responsible and environmentally sound clothing, and search the internet for the most eco-conscious brands.

AVOID LEATHER: leather is not just a by-product of the land-hungry meat industry, it promotes it by being a big earner, and the softer leathers can come from unborn animals cut from their mothers' wombs. Leather production requires large amounts of water and energy, and the leather-tanning process is among the most toxic in all of the fashion supply chain, causing carcinogenic chromium to be released into the water table. It's well known that leather-tanners have an increased risk of cancer. If you must buy leather, look for vegetable-dyed brands.

BUY SUSTAINABLE SHOES: shoes can be made with some surprising materials, including pineapple leather and recycled car airbags. They can also be made from cork, sustainable leather, hemp, recycled polyester, wool, coir, natural rubber, banana leaves and recycled tyres.

RECYCLE TRAINERS/SNEAKERS: schemes such as the Nike Grind programme turn old sneakers into a material used in footwear, sports tracks, courts and playgrounds. They accept any brand of athletic shoes – drop them into your nearest Nike store for recycling. Just do it.

Save the World When You...Use the Bathroom

One of my favourite things in the world is to sit next to a bubbling spring. Something about being there when water that might not have seen daylight for 5000 years returns to the Earth's surface feels kind of magical. But, more than 95 per cent of Earth's freshwater (not including ice) is stored in underground aquifers and this groundwater is being used far more quickly than it is being replenished. As the population rises we are, literally, draining the life force that pumps through our planet. So much so that some cities around the world are even starting to sink from the volume of groundwater being pumped out from under them.[1]

So, how about we become water stewards, making good use of the water we have by not wasting it or polluting it with toxic chemicals and microplastics, and doing what we can to preserve and protect what's left? In the UK, a third of water taken from the natural environment is wasted through leaks, at treatment works and in the home.[2] And, as personal washing accounts for around 33 per cent of the water used in the home, let's start by cleaning up our act.

SINGING IN THE SHOWER

Hot water is the second largest consumer of domestic energy after heating our homes. We spend on average around eight minutes in the shower each day,[3] which translates to around 90 litres (24 US gallons) of water down the drain. If we could just reduce that to five minutes, we'd save a ton of water, energy and money. An easy way to change your showering habits is to aim for your shower to last no longer than it takes you to sing a song. If you need to wash your hair a couple of times a week then you could probably get away with singing it twice. And, as a bonus, you'll have something to share with your friends next time you're sitting around a campfire from all that practice.

WATER-SAVING SHOWERHEADS

If your water supply is metered, these will save you money as well as reducing your water consumption. Here's a mind-blowing statistic for you: if every American installed WaterSense-labelled showerheads it would generate annual savings of over $1.5 billion (that's about £1.15 billion) in water utility bills, more than 250 billion gallons (approx. 950 billion litres) of water and around $2.5 billion (£1.9 billion) in energy costs for heating water.[4] So not only will you be better off financially, your conscience will be as squeaky clean as your freshly-showered ass.

BATH BUDDIES

Sharing a bath with someone you love, or at least like a lot, has got to be one of the most rewarding world-saving actions you could take. And, if you don't want to get naked with your flatmate, or your kid doesn't want to see you naked, you can always take it in turns. Unless said flatmate (or kid) has just returned from, say, a festival. As with any rule, there will always be exceptions.

It's obviously not the most impactful action you can take to save the world, but goddammit, saving the world isn't always easy and sometimes you might just need a long, hot soak and a back-rub. And, if you do this instead of watching TV and your water is solar-heated, then you're really talking earth-positive tub time. And, don't forget, you can leave the water in the tub and use it to water your plants, especially if you had Epsom salts in there for your aching muscles – they make a great homemade plant-food.

BEAT THE BEAD

Microbeads are tiny plastic spheres used in facial washes, shower gels and even toothpaste. Being so small, they pass through sewage filters into our waterways and into our oceans. And they look, to hungry sea creatures, exactly like delicious fish eggs. Microbeads and other microplastics – tiny particles of plastic from paint, car tyres and synthetic clothing – are forming a great toxic plastic soup in our oceans. Have a read of the microfibres section of the 'Save the World When You...Dress' chapter for more on why this is bad news for marine life, as well as seafood-eaters. The simplest way to avoid microbeads in your bathroom products is to choose organic products using ingredients such as apricot kernels, or to make your own facial scrubs. If a product is labelled 'organic' it won't contain ocean-polluting microbeads. If you prefer to make your own, sugar is a fine exfoliant, as are used coffee grinds.

Fortunately, the UK, France, New Zealand, Taiwan, Canada and parts of the US have banned microbeads in personal-care products, but it's still worth looking at the label, as some brands have already started switching to bioplastic, which as we talked about in the straw section of 'Save the World When You...Drink', could well cause as many problems as its plastic counterpart. Also, some forms of microplastics have yet to be banned, such as the ones found in makeup and sunscreen. Yep, they can have plastic in them too. The simple answer? Buy less of them and, with the money you save, buy organic versions of the ones you can't live without.

TOOTH FAIRIES

Over the course of your lifetime, you're likely to get through around 300 toothbrushes. And unless you use them for scrubbing sinks, football boots or brushing your hamster (yep, that's a thing), they're unlikely to get reused, and the vast majority of them will end up in landfill, slowly fragmenting

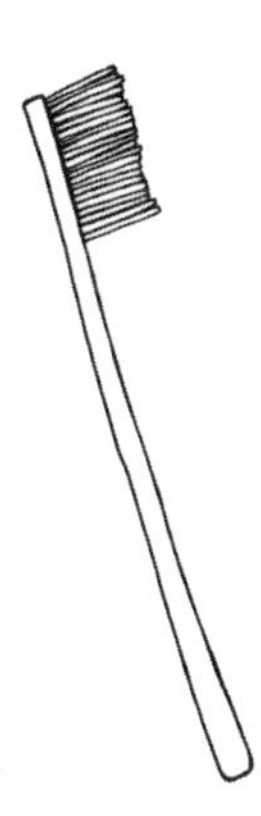

into pieces, leaching toxic chemicals into the earth for hundreds of years to come.

Figures for the number of plastic toothbrushes sold worldwide every year range from 3.5 billion to 4.7 billion and, somehow, either through ineffective waste-management systems, fallen cargo containing plastic waste, or from open landfill sites next to rivers, a whole heap of toothbrushes make their way into the ocean. They've been found inside the bellies of seabirds and are a frequent find during beachcleans around the world.

Enter the bamboo toothbrush! They're abundant on Instagram and they're every budding environmentalist's essential piece of plastic-free kit. Well, some of them are plastic-free and fully compostable and biodegradable, but those ones usually have some kind of animal bristles. If you're vegan and would rather go swine-free while you clean your teeth, then switching to a bamboo toothbrush with nylon bristles is better than an entirely plastic toothbrush. You can pluck nylon bristles out of the toothbrush head when you're done with it and put them in the bin. Weighing in at less than 0.3 grams, that's 98 per cent less plastic going to waste than from an entire toothbrush. You can repurpose the bamboo handle afterwards when making homemade ice-lollies or as plant markers in your blossoming veggie patch. Now, there's something to smile about.

THE THREE 'P'S – PEE, PAPER AND POO

More than once I've been introduced on stage as 'the amazing poop-talker'. No, that doesn't mean I talk shit! It's because I've ended up becoming a bit of an expert on plastics that get flushed down the toilet. And an advocate for all things natural that should go down the loo, like pee, paper and poo (or 'poop').

Water companies in the UK spend around £100 million a year unblocking sewers (about US$130 million), an avoidable cost if we all stopped flushing things like wet wipes, menstrual products and cotton buds (QTips), down the loo.[5] The problem, however, doesn't just lie in the cost to our water companies (and therefore to those of us who are paying the bills) but in the cost to the oceans. Almost a tenth of plastic found on UK beaches is coming directly from our toilets, according to the Marine Conservation Society's 2017 Beachwatch report.

FATBERGS

Perhaps don't read this next bit if you're on your lunch break. When all of the aforementioned 'unflushables' get into our sewers – along with all the fats, oils and greases which are put down the sink but should go into your compost or waste bins – great big greasy, stinky clumps called 'fatbergs' form and block the sewers. Another revolting outcome of flushing anything other than the three 'p's down is that, during heavy rainfall, combined sewage systems can't cope with the extra volumes of water and have to open the floodgates, literally, into our rivers and seas. These Combined Sewage Overflows (CSOs) discharge everything that we've flushed down the toilet – all the wet wipes made of plastic, the tampons and pads – to bypass sewage filters and head for the open sea.

In 2018, London-based conservation group, Thames 21, found over 5000 wet wipes in 100 square metre patch of Thames riverbank (just over 1000 square feet). It's not pretty. And all the squeaky-clean bumholes in the world aren't worth that kind of environmental pollution.

So, only flush the three 'p's – pee, paper and poo. Anything else goes in the bin. Make sure you have a bin in your bathroom, or two if you want to separate out your wipes from your recycling. And, if you really must have that fresher-than-fresh feeling after you go to the loo, invest in a set of washable cheeky wipes, or a bidet hand shower, or just get a bidet on eBay. For people with periods, read on.

MORE WAYS TO SAVE THE WORLD WHEN YOU USE THE BATHROOM

GET FREE WATER-SAVING DEVICES: toilet-cistern bags, tap inserts and bathwater diverters as well as eco showerheads, shower regulators, shower timers (to tell you when to get out) could be available free from your water provider. A ridiculously fast and easy way to check this, if you're in the UK, is to go to www.savewatersavemoney.co.uk and enter your postcode, then pick your device. For US readers, get in touch with your local water supplier to see what free water-saving devices they offer.

SWITCH TO PLASTIC-FREE TOILETRIES: delightful and efficient non-plastic toiletries are out there in abundance. Try non-disposable metal or bamboo safety razors, bars of soap, shampoo bars, natural toothpaste in glass jars, bamboo dental floss in a glass jar, shaving soap bars, deodorant crystals, paper and bamboo cotton buds (QTips), and more.

USE NATURAL AND BIODEGRADABLE TOILETRIES, PERFUMES AND COSMETICS: the average British woman comes into contact with 680 'chemicals of concern' through toiletries every month, and these get washed into our water supply, polluting rivers and seas as well as our bodies.[6]

MAKE YOUR OWN ECO-TOILETRIES AND ROOM SPRAYS: this is also a good activity for kids and teenagers; the internet is full of fabulous ideas, such as making your own toothpaste from bicarb, coconut oil and peppermint oil – you can use it as an effective deodorant too.

AVOID ANTIBACTERIAL HAND-PUMP SOAP: apart from creating vast amounts of polluting plastics, these often contain Triclosan, an endocrine disruptor that will soak into skin and enter the water system. Triclosan also reacts with sunlight to create dioxin, a highly toxic carcinogen, and has been blamed for antibiotic resistance.

USE WASHABLE MAKEUP REMOVER PADS: these dispense with the need for cleansers and cotton pads – add water, rinse and repeat!

IF IT'S YELLOW, LET IT MELLOW; IF IT'S BROWN, FLUSH IT DOWN: avoid unnecessary flushing and, if you don't have a high-efficiency toilet, use the time-tested method of putting a brick in your cistern to save masses of water when you flush.

KEEP TWO BINS IN YOUR BATHROOM: one for recyclables, the other for wipes, menstrual products and the hair you pull out of the plughole. This way your recyclables don't get messy and it makes the sorting all the more hygienic.

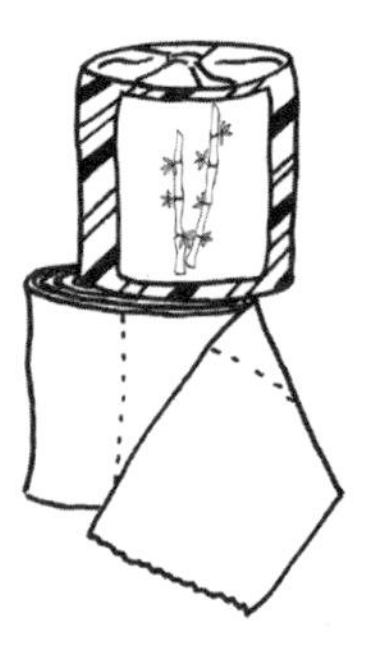

USE ECO-FRIENDLY TOILET PAPER: if every household in the US replaced just one roll of virgin fibre toilet paper (500 sheets) with 100 per cent recycled ones, we could save 423,900 trees a year.[7] But don't just buy one, switch to unbleached, recycled toilet paper and save millions more trees.

USE ORGANIC, NATURAL-FIBRE TOWELS: polyester ones release huge amounts of microfibres during washing.

BUY A NATURAL SHOWER CURTAIN OR INSTALL A GLASS SHOWER BARRIER: PVC shower curtains contain toxic chemicals which will contaminate your home and leach into landfill sites.

SHOWER HOT THEN COLD: start off with a warm shower, then turn the temperature right down. Cold showering saves energy, improves circulation, boosts immunity, can relieve depression and aid weight loss, and is less drying on skin and hair, since hot water depletes our natural oils.[8]

Save the World When You...Bleed

Close to 20 billion menstrual pads, tampons and applicators are dumped into North American landfills every year. And, in the UK, around 2.5 million tampons, 1.4 million pads and 700,000 panty liners get flushed into sewers *every single day*.[1] This means Brits are collectively flushing the equivalent of 5.6 million plastic bags down the loo every day, since pads can be made of up to 90 per cent plastic, and most contain the same amount of plastic as four carrier bags.[2]

I'd love for anyone who doesn't have periods to read this section too, as I imagine you may find this useful. And, for the love of the oceans, get a bin in your bathroom so visiting friends and family can stick their tampons and pads in it when they come to visit. If you're a granny, you may enjoy reading how far things have come since the sanitary belt and napkin. And if your grandkids don't know what a Mooncup is, you can be the one to tell them.

The period-product marketplace hasn't been this exciting since the invention of the tampon. From washable pads to menstrual cups, panties you can bleed into, reusable applicators and a range of organic disposables, there's never been a better time to make your red flow green. And, yes, with the arguable exception of organic disposable pads, they all save you money.

PLASTIC-FREE PERIODS

If you've just read the previous section on flushed plastics, you'll be well aware that flushing high-plastic menstrual products down the loo causes big problems for us and the fish. Pads are full of plastic, and tampons, whilst predominantly cotton (which, unless organic, is still pretty disastrous for the planet) also have components made of polyester.

The average menstruator will use between 11,000 and 15,000 disposable menstrual products in their lifetime and, given that plastic won't actually biodegrade, that adds up to a serious amount of waste.[3] With around 2 million items of period waste littering the whole of just the UK coastline[4], it's time to turn the tide.

WASHABLE PADS

If home-stitching yourself some stylish washable pads isn't your thing (I'm a terrible seamstress, my homemade menstrual pads wouldn't hold a sneeze-wee), then have a look online to get some from someone who knows what they're doing. One person in the UK has been getting a lot of practice lately: Bryony Farmer started her company, Precious Stars, in 2013 at the age of 15. In her first six months of trading, she sold £2,000 (around US$2,600) worth of homemade, hand-stitched reusable pads. Fast-forward to 2018 – over 100,000 YouTube subscribers later – and she's selling around £43,000 worth of

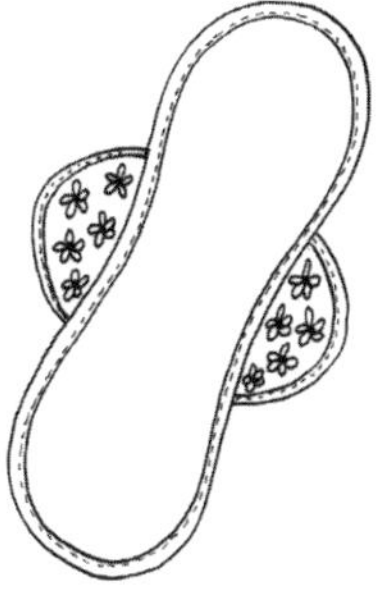

pads (that's over US$56,000) in the same time period (she's obviously not making them all herself now and her parents are quite pleased to have their dining-room table back!)

PERIOD PANTIES

This is another ingenious invention rocking the world of periods. Thinx were first on the market with their absorbent underwear designed to replace, or complement, tampons and pads. In money-saving terms, given the average menstruator pays around £5 (up to US$7) a month for pads and tampons, it would take around 15 months for a set of three Thinx to start saving you money. But, since Thinx first came on the scene, more producers have turned up offering cheaper versions. If you want to try them out, look for organic cotton or bamboo brands.

DISPOSABLE PADS

Organic menstrual pads are definitely worth a mention. Yes, they cost a bit more than their plastic-laden, bleached and soil-sapping, non-organic cotton alternatives – on average you'll pay around seven pence (nine cents) more per pad. But, aside from the environmental benefits, you might want to consider switching for health reasons too. In 2014, an analysis of various Always-branded pads by Women's Voices for the Earth, the environmental research organization, found they contained traces of 'styrene (a human carcinogen), chloromethane (a reproductive toxicant), chloroethane (a carcinogen), chloroform (a carcinogen, reproductive toxicant, and neurotoxin) and acetone (an irritant)'.

Brands such as Natracare are worth championing: they've been leading the way in organic menstrual products for 30 years, looking after the skin we're in and the planet we're on.

MENSTRUAL CUPS

At City to Sea, the organization I founded back in 2015 to stop plastic polluting our rivers and seas, we coined the phrase 'the rise of the reusable' when championing the success of products such as menstrual cups, which are a fantastic solution to plastic pollution from our periods. The rise in popularity of the reusable cup (known to many by the brand name Mooncup) is definitely happening; the number of Google searches for 'menstrual cup' increased by almost 300 per cent from 2013 to 2018.[5]

If you've not heard about them, here's the 'flow-down'. A menstrual cup is essentially a soft little silicone bucket that is inserted into the vagina during menstruation to catch the blood. When it's full, or after a number of hours, you gently tug it out, empty it into a toilet or sink, rinse it and reinsert. At the end of a period it gets washed and sterilized and sits happily on the shelf until next month. And so on, for several years. Unless it gets eaten by your dog when you accidentally leave it sitting on the edge of the bathtub, as happened with mine. Oops.

Menstruators the world over are discovering the benefits of switching to menstrual cups for more than the environmental benefits too. A switch from conventional disposables to a menstrual cup can save you around £77 (US$100) a year.

REUSABLE APPLICATORS

Here's one to watch. UK-based start-up, Dame, recently crowdfunded over £100,000 (more than US$130,000) to create 'D', the world's first reusable tampon applicator. It uses self-cleaning technology and medical-grade materials – meaning that those who prefer the smooth glide of an applicator can do so without the plastic waste.

ORGANIC TAMPONS

You should be clued up on the environmental reasons to use only organic cotton from now on, and that goes for tampons, too. The average menstruator will have a tampon inside the vagina for more than 100,000 hours over her lifetime[6] and yet the US Food and Drug Administration (FDA), for example, does not require companies to list the ingredients. Tampons can contain carcinogenic pesticide residues, phthalates, azo dyes and dioxins, which, although present in low levels, could theoretically increase your risk of disease after 100,000 hours of use, in one of your most sensitive and absorptive body parts![7] And it won't cost you more either: non-applicator Lil-Let tampons cost the same as Natracare's organic equivalent.

MORE WAYS TO SAVE THE WORLD WHEN YOU BLEED

SHARE VIDEOS AND CONTENT ON SOCIAL MEDIA ABOUT 'THE RISE OF THE REUSABLE': help spread the word about reusable menstrual products to menstruators, parents and teachers around the world.

REMEMBER THE THREE P'S: only pee, paper and poo go down the loo, so don't flush pads, panty liners or tampons!

SWITCH TO REUSABLE TAMPONS: these are made with organic cotton and are designed to be washed after every use.

INVEST IN SOME PERIOD PANTIES: do away with panty liners by having a couple of pairs of period panties for light-flow days.

TRY A REUSABLE MENSTRUAL DISC: an ingenious invention, such as the Ziggy Cup, that some menstruators prefer to a menstrual cup as they catch the flow further upstream than cups. They can last twelve hours, hold the same amount of blood as five tampons and, users claim, they can have leak-free sex while using them.

USE WASHABLE PADS MADE FROM BAMBOO: for an ultra-eco, zero-waste period, make sure your cloth pads are made from sustainable fibres.

LOOK INTO USING A SEA SPONGE INSTEAD OF A MENSTRUAL CUP: used carefully, with proper sterilization, sponges can be a great natural, reusable product for periods. While more prone to leaking than menstrual cups, users say they can be more comfortable.

BALANCE YOUR HORMONES: if your periods are heavy, you could consider supplementing your diet with calcium for two weeks before your period or exploring herbal supplements to help balance your hormones.[8] A lighter period has the bonus of being better for the environment too – the lighter your flow, the fewer period products you'll need to use.

Save the World When You... Exercise

Saving the world takes energy, as anyone who studied even the smallest bit of physics will tell you. If you want the status of something to change, some other factor has to enter the equation. When it comes to changing the world for the better, you are that other factor. So you're going to need to look after yourself, know how to top up your tanks and how to maintain a positive mental attitude. You need to know not just how to survive, but how to *thrive*.

This is where getting active comes in. Exercise helps keep us strong and energetic. It also releases endorphins, the feelgood chemicals that relieve stress and enhance pleasure. But what if there was a kind of exercise that not only ticked the endorphin box, but that of four other factors that make us happy? Well, there is! Read on for more.

GOOD GYM

Good Gym is one of my hands-down favourite ways to save the world for free. Set up by a group of runners in the UK who believe gyms to be a waste of energy and human potential, Good Gym aims to redirect that energy towards neglected tasks and people in our communities. Basically, you meet up with a bunch of people, typically lovely types who will enrich your life, run to a community project that needs some extra pairs of hands, do said mission for around 45 minutes, then run back. For free. Although you're invited to donate £9.95

(US$13) a month if you want to, but that's optional and no one will judge you if you don't!

There are five scientifically proven actions that have been shown to improve personal wellbeing.[1] And going on a Good Gym group run ticks them all: you *connect* with people by running in a group, often with people that you've not met before or wouldn't meet in your day-to-day life; you're *active* through running and doing something physical for your task; you *take notice* of new routes through your town or city, spotting wildlife and changing seasons; you *keep learning* from each other, whether gardening skills, running techniques, wild food or by learning about the project you're supporting. And you *give*, by doing something nice for someone, just because you can. And this all happens in under 90 minutes, with the added option of a refreshing local beer at the end if you choose!

THE PERKS OF PARKRUN

Another free and fabulous way to get fit and stay active is to sign up to Parkrun. This organization runs free, weekly, 5km (a little over 3 miles) timed runs all around the world. They're open to everyone and they're friendly and easy to take part in. If Saturday mornings sound like your ideal workout time, have a look online to find one near you – they're currently happening in 20 countries across five continents.

PEDAL POWER

Ever thought that all those calories burned in a spin class could be harnessed? Or that all the miles clocked up on treadmills could be generating electricity? Well, some smart cookies around the world have worked it out and you may just have a 'human-powered' gym in your home town. Instead of all that energy you're generating in your workout just vanishing, it actually powers the very machines over which you're working up a sweat and any excess energy is sold back to the grid.[2]

ONLINE WORKOUTS

If you want to save money and don't need the motivation of a gym membership to get you working out, try having a look online for free workout classes that fit into your schedule. From yoga to boxercise, you can do away with the travel to and from the gym, fancy gym kit and monthly fees, and get fit in your own front room.

REDUCE YOUR CARBON PAW PRINT

If exercise for you means walking your dog, then you, as well as all those disgruntled pedestrians with dirty shoes, will know that it also means dealing with dog poo. A million tonnes of dog poo is produced every year in the UK, and over ten times that in the US – and then there's the litter from tens of millions of cats. That's quite a poo print to leave on the planet, whether it's picked up and put in a plastic bag – and sent to landfill – or not ('not' being the case with about half of the poo). And it's not just the volume – it's what's in it that's an issue. Dog poo is an environmental pollutant classified alongside oil and toxic chemicals, in the US, by the Environmental Protection Agency (EPA).

WHAT TO DO WITH DOGGY DO

As things stand, the best option is the 'stick and flick' method – but that won't go down well if you're walking your dog in an urban area or a playground. For city-dwelling dog owners, using biodegradable bags is the next best option, although don't be fooled by 'oxo degradable' or even some 'oxo biodegradable' bags which do not break down completely. Oxo degradable plastics just fragment into smaller particles faster than conventional plastic so best to buy corn starch bags instead. If you're in the UK, don't be tempted to flush dog poo down the toilet, as it is processed differently to human waste and can be toxic to workers. In the US, however, flushing is recommended by the EPA. It's also fine to bury dog poo, so long as it's away from water sources or your vegetable patch, and it can be composted at home using specialized methods (but don't add it to your usual compost). And, of course, if you're in the countryside there's no reason to use a bag at all – the trusted 'stick and flick' method is the most environmentally friendly way to deal with the do.

FUEL FROM STOOL

Some eco-minded innovators across the globe have been building gas lamps powered by methane biofuel from dog poo, with, for example, Matthew Mazotta's continuous-flame gas lamp in Cambridge, Massachusetts, and Brian Harper's gas lamp in the UK.[3] This is a brilliant way to reduce both methane production and toxic compounds from dog poo, as well as the public expense of disposing of dog poo – while making those winter dog walks all the brighter.

MORE WAYS TO SAVE THE WORLD WHEN YOU EXERCISE

WALK OR CYCLE THERE: try to fit more walking or cycling into your normal routine by taking short journeys by foot or bicycle or – if you have to drive – try parking further away from work or home and walking the last bit.

TRY THE COUCH TO 5K PROGRAMME: this is a great, free way to get started on your fitness regime and operates all over the world. There's also a free app.

GO PLOGGING OR TRASHERCIZING: this entails going out for a hike or jog and picking litter up along the way, which means you add bending and squatting to the mix. It also burns 288 calories compared to 235 on an ordinary run or jog.[4]

DO A BEACHCLEAN: this is a great way to educate kids about protecting our beaches, and all that squatting to pick things up is good for flexibility. Check the internet for beachclean events near you or download the #2minutebeachclean app and get connected to your beachclean community!

TAKE THE STAIRS, NOT THE LIFT: footballers are trained to reach peak fitness by running up and down stadium steps and there's no reason why you can't, too. It's free and you won't waste electricity by making the lift go up and down.

GET FREE EQUIPMENT: look at online reuse sites such as Freecycle for free skipping ropes, exercise balls, weights, or even a bicycle.

SWITCH TO ENVIRONMENTALLY FRIENDLY WORKOUT CLOTHES: if you're looking for new workout gear, source it second-hand or buy kit made from hemp or bamboo.

USE NON-TOXIC, ECO-FRIENDLY YOGA MATS: made from cork, jute or natural rubber.

REFILL IT: don't forget to take your reusable water bottle to the gym or whenever you go out exercising.

USE FREE GYM EQUIPMENT IN PARKS: lots of parks have Green Gyms or all-weather workout equipment for anyone to use. Who cares about getting red-faced in public when you're high on endorphins?

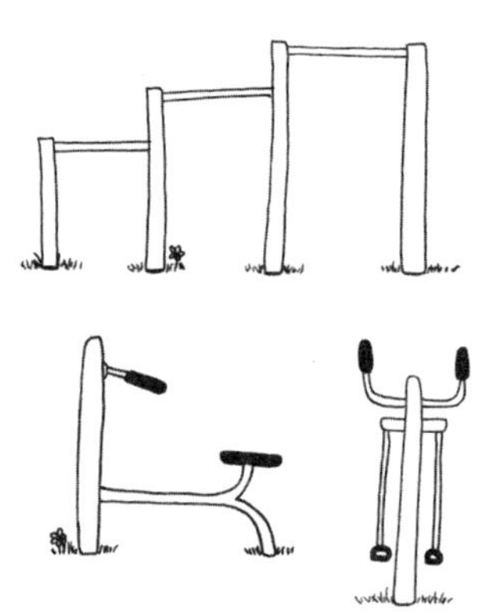

Save the World When You... Have Sex

Yes, really. Considering that sex is one of our favourite pastimes, eco-sex doesn't get a lot of action. So, this chapter is dedicated to making our lovemaking even more pleasurable, safe in the knowledge that we're not screwing the planet. From avoiding unwanted pregnancies to knowing what to do with your vibrator when you're done with it, engaging in earth-positive sex can be a deeply satisfying experience for lovers and loners alike. I'm sure Marvin Gaye would have agreed that, with the world getting 'hot just like an oven', it needs some lovin'. And, when it gets that feeling, we need eco-sexual healing...

WE'RE GOING TO NEED A BIGGER BOAT

It's time to touch on the biggest taboo when talking about saving the world...the unmentionable, enfant terrible of the environmental movement. Population. According to a study at Lund University, Sweden, the single biggest action you can take to save the world is to have fewer kids.[1] While baby-making with someone you love, and raising a child, can be one of the sweetest things on earth, drawing the line – or snipping the tubes – at two is the planet-friendly thing to do. And, since we're talking about saving the world for free, it's good to note that in 2018, the average cost of raising a child to the age of 21 in the UK is £231,843 (over US$300,000).[2] You could self-build a wonderful eco home for that.

GIRL POWER

In addition to your own choices around baby-making, improving access to education for women is another way to address the population crisis. All around the world, surveys have shown that girls who get a secondary school education have fewer kids, and have them later in life.[3] One way you can help, for free, is to find a charity focused on providing education and access to contraception for women, and share their content far and wide.

GREENER BIRTH CONTROL

If having children is 100 per cent part of your life plan, and you're fortunate enough to be able to have them, then go forth and be a magnificent parent. And, until such time as you consciously choose to procreate, let's explore some of the greener birth-control methods available to everyone who wants to do their bit while getting it on.

CONDOMS

At this point we're going to do away with the 'if you can't reuse it, refuse it' rule. In the greater scheme of reducing emissions by not getting yourself or someone else pregnant, a pile of single-use condoms and wrappers is worth the trade-off. When it comes to

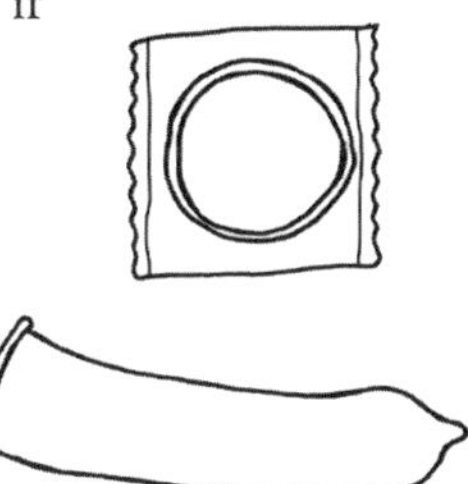

casual sex, condoms are king and can be found in planet-friendly varieties, such as biodegradable lambskin, Fair Trade rubber or vegan condoms (the latter are easy to source online and won't cost you more than conventional condoms, which contain casein, a milk derivative). Just don't flush them down the loo when you're done with them... everrrr. (See the chapter 'Save the World When You... Use the Bathroom' on p.136 for a reminder of why.)

IUDS (INTRAUTERINE DEVICE)

Copper, non-hormone-based IUDs (or coils), seem to be the most popular choice for women who want a zero-waste birth-control method but also want more reliability than the fertility-awareness methods that we'll talk about in just a mo. The copper T-shaped coil, once implanted, can give you up to 12 years of protection (from getting pregnant, not from STIs), and is super-reliable. They're not for everyone though, as some women report heavier periods (while many experience lighter ones), and some wombs decide they don't want a foreign object sitting in there and reject it.

CALENDAR-BASED METHODS

Our bodies are very clever and, with a healthy diet and enough exercise, tend to run like clockwork. If we get to know our rhythms well enough, with our partner's co-operation we can avoid getting pregnant. Fertility-awareness methods, or 'rhythm' methods, are naturally zero-waste as a means of contraception and are increasing in popularity – perhaps thanks to the rise of some very nifty apps on the market (some of which are free). This method does involve diligence in terms of monitoring your cycle and temperatures, and the failure rate is high, so it's not for everyone.

THE 'PILL'

The most widely used birth-control method and, therefore, the tops in terms of helping to stem the flow of us car-driving, shopaholic, meat-eating humans. So, as far as carbon emissions are concerned, the pill is very much worth the waste it creates. However, women on the pill are peeing out the synthetic oestrogen ethinyl-estradiol (EE2), which ends up in our rivers and seas.[4] And, with around 100 million women worldwide on the pill, that's one tsunami of hormones flooding our waterways with the inevitable negative consequences for fish and us humans. Multiple studies have shown that EE2 can alter both behaviour and genetic balance in freshwater fish, causing fish to develop problems with procreation, which can lead to the complete disappearance of an entire fish population.[5] You could try peeing in a jar and pouring it over your compost heap instead of flushing it down the loo (the nitrogen helps the composting process) but that might not be for you...especially if you don't have a compost heap. One hundred million women on the pill[6] also means 1.2 billion pieces of non-recyclable plastic packaging being discarded every year. One way to offset the waste is to try to influence the people that make them. Send a letter, email or tweet to the manufacturer of your brand of contraceptive pill and ask them to use aluminium blister packs, which are recyclable. The millions of women around the world using Hormone Replacement Therapy (HRT) could do this world-saving action too.

DON'T F*CK THE OCEANS

Whether you're entertaining a guest or entertaining yourself, knowing how to give and receive pleasure between the sheets is an essential skill for anyone committed to saving the world. For starters, it's fun, it's good for you and it's free. When faced with sizeable challenges (such as saving the world from irrevocable climate change within the next decade) having a few healthy ways to de-stress and unwind will go a long way to ensuring you don't burn out. But given this chapter is about sex and not about going on long walks, let's make sure any 'extras' we bring into the bedroom get Mama Nature's juices flowing as well as ours.

GLASS IS CLASS

The sex-toy business is huge – over £11.5 billion a year huge (that's US$15 billion). Some 60 million sex toys are sold globally every year[7]...but have you ever seen one at the local recycling centre? Neither have I. I've seen more than one on a beach clean though. What to do? Well, when it comes to the world of ethical sex toys, glass is a pants-down winner. Rubber-based or plastic dildos, plugs and beads can contain phthalates, a seriously dodgy cancer-causing chemical that you really don't want messing with your membranes. Glass dildos, on the other hand, are not only cheaper but they're made from high-grade borosilicate glass, the same sturdy material that is used in pyrex jugs and oven dishes. And, if you're sensitive to latex or rubber, you'll be glad to know glass won't give you any trouble. You can heat glass toys up, cool them down, wash them off, share them with your partner (after the washing-them-off part), slather them with lube... all without damaging them or risking them deteriorating and leaching toxins into your most intimate of areas.

GOOD VIBRATIONS

If vibrators are your thing, opt for phthalate-free, rechargeable ones (you've switched to a green-energy supplier, right?) and, if you're shopping online, make sure you send a note to the seller asking for it to be shipped plastic-free. There are some carbon-neutral sex toys on the market too, which may have a higher price tag but whose quality should ensure you have a lifetime of fun. You can even have solar-powered orgasms. How's that for hot?!

BUZZTED

As for recycling your sex toys, if you live in the UK and put your vibrator in the bin, you could actually be breaking the law. Laws about what can and must be recycled differ between countries, but in the UK, the Waste Electronic and Electrical Equipment (WEEE) regulations mandate that households recycle electrical items. So, next time you're visiting your household-waste recycling centre, make sure to take any broken vibrators with you. It's about time some of those 60 million sex toys got a new life anyway.

PLANET-LOVING LUBE

Here's a fun one. Well, they're all fun really, but this one's extra fun. Why, oh why, buy processed lube in a single-use tube when you can delight in nature's slippery goodness for a fraction of the cost? Here are three simple-yet-sexy, cheaper-than-off-the-shelf lube solutions for you to experiment with.

ALOE VERA

While this isn't really suitable for oral play (unless that bitter taste drives you wild), a freshly cut slice of aloe will give you an ample supply of natural, slimy goodness to help you out of (or into) a tight spot. Working this closely with nature is a turn-on in itself and will leave your pleasure-parts soothed and supple.

COCONUT OIL

This is a bedroom essential; great for massages and great for happy endings. Coconut oil usually comes in glass jars which can be recycled or used to store your dry nuts, perfect for your after-sex snacks, you frisky little squirrel. Don't use it with latex condoms though – oil-based lubricants can damage latex and cause the condom to break.

FLAXSEED GEL

This one is my favourite, not least because it doubles up as a hair-styling product. Get yourself a bag of loose flax seeds from your local scoop shop or health-food store. Add two tablespoons of flaxseed to two cups of water, bring to the boil, simmer for about 20 minutes or until the water has reduced by half, strain off the remaining liquid and, very importantly, let it cool. It'll keep in the fridge for a couple of weeks or two days unrefrigerated. Give it go, it's totally amazing and not as seedy as it sounds.

PLANET-POSITIVE PORN

It's perhaps an unusual topic to cover when talking about saving the world but, given that just under one third of women[8] and three quarters of men[9] watch porn online at least once a week, I thought it prudent to see if we can start greening this bigtime pastime of ours.

Very little has been written on the carbon footprint of online porn, despite porn sites purportedly receiving more regular traffic than Netflix, Amazon and Twitter combined each month.[10] The internet is responsible for around 300 million tonnes of carbon dioxide per year, which is the same amount as every person in the UK flying to America and back, twice.[11] Given some sources claim that up to 30 per cent of internet traffic is porn-related, it's looking like porngasms come with a hefty smack on the planet's cheeks.

But first let's differentiate between ethical porn, which is a thing, and eco porn, which isn't. Yet. Ethical porn is a growing sector of the porn industry, but it's not free, as most truly ethical porn sites (the ones that give their actors equal pay, provide health checks, offer reasonable working hours and have clear consent processes) involve you paying either a one-off fee or a subscription to watch it. Which is fair enough and probably one of the simplest ways to make your porn viewing more mindful in terms of the wellbeing and safety of those involved.

In terms of the huge amount of data used to stream porn every day around the world, and the energy that consumes, maybe planet-loving porn users need to find a few offline ways to get their kicks. Below are some suggestions to get you going.

MORE WAYS TO SAVE THE WORLD WHEN YOU HAVE SEX

LEARN TO FIRE UP FROM FOREPLAY INSTEAD OF ONLINE PORN: if you're having sex with someone other than yourself, find some screen-free ways to get each other turned on while the screens stay turned off.

GIVE GREAT HEAD...MASSAGES: find a free course or do a skills-swap with a friend to learn new ways of pleasuring the body through massage.

SOURCE SOME SECOND-HAND EROTIC LITERATURE: reading it to yourself or taking it in turns with your partner can be a fun, eco-friendly way to get aroused and learn new techniques.

COMPOST YOUR CONDOMS: if they're made of natural latex, they're biodegradable, although they take a while. If you don't fancy composting them, throw them away – just don't flush them down the toilet.

SKIP SHOWER SEX: unless you've got it boiled down to four minutes, which is the recommended amount of eco-time we have the water on for. Go ahead and share the shower afterwards, though.

TRY VEGAN APHRODISIACS: instead of oysters, try eating beetroot, watermelon, strawberries, leafy greens and/or pistachios, as nutrients in them boost circulation and libido.

WEAR PLANET-LOVING LINGERIE: there are lots of ethical, sustainable lingerie brands out there using, for example, fabric-production leftovers and vintage trimmings, Fair Trade organic cotton, cruelty-free silk, recycled water bottles and organic hemp and bamboo.

AVOID CHEAP VALENTINE'S DAY GIFTS: stay clear of fluffy toys and disposable presents. Instead, go for a homemade card, cellophane-free flowers, ethical eco-jewellery, dinner for two at a zero-waste restaurant, or tried-and-tested organic chocolate.

GREEN-LIGHT THE CANDLES: turn off the lights and burn some air-purifying beeswax or soy-wax candles, which don't contain paraffin. Some even have wooden wicks, which prevents the lead contamination which can be caused by the ordinary type.

TRY AN ECO-FRIENDLY MINI-BREAK: instead of jetting off for a weekend city break, try staying local in a romantic shepherd's hut, yurt or treehouse (ideally with a solar-powered hot tub, of course), getting there by train or bike to prevent bickering in traffic jams.

DO SOME ECO-COUPLE ACTIVITIES: instead of going to the cinema or pub, try visiting a wildlife reserve, doing conservation volunteer work together, visiting an organic winery, going stargazing, or cooking a healthy meal with ingredients from the local farmers' market to celebrate saving the world together.

Save the World When You...Play

Other than having sex, which we've covered in the previous chapter, there are lots of ways to have fun, relax and recharge that don't cost the earth. Knowing a number of ways to feel good that don't involve shopping or driving your car is an essential skill for anyone who wants to save the world for free. There are rich and varied experiences on our doorsteps just waiting for us to discover them. Learning how to get your kicks for free in a way that doesn't burn through the earth's finite resources often leads you to doing things that are really good for you. Like singing, dancing, sharing stories, growing things, learning stuff and having some much-needed time away from a screen.

ME TIME

Sometimes, spending a little quality time with 'me, myself and I' can be just the ticket for anyone that needs a little time out, free from chatter and to-do lists. And if you want to save the world while you do it, here are some ways to play that will recharge your batteries and keep you energized as sustainably as a Tesla Powerwall...only a lot cheaper.

HEAD OUTSIDE

Nature is waiting for you. It's what I'm writing this book for and the reason you have it in your hands. So, get out there and make the most of it. Seek out your local wild spaces, visit an ancient spring or well, or hop on a train to your nearest mountain or ocean. Run, walk, wild-swim, cycle, whatever... just go.

MAKE MUSIC

Having musical instruments around the home is a wonderful invitation to play. I can't really play the piano (those who know me will know I can play one song, badly) but getting one for my home was one of the best things I ever did. I got a cheap one on Gumtree and now, whenever anyone musical comes over, it gets played and we have fun. Other than a guitar, which everyone should have in their house, another cool instrument to own, which literally anyone can play, is a shruti box, which creates a basic Indian drone that even the world's least musical person can have fun with.

SPEND TIME IN YOUR GARDEN OR ALLOTMENT

I know I wrote a whole section on gardening, but it's so good for you and so good for the bees, I thought I'd slip it in again. It's almost impossible to worry when you're gardening. And you get to eat the fruits of your labour too.

WRITE A LETTER OF APPRECIATION TO SOMEONE YOU LOVE OR ADMIRE

Instead of getting a dopamine rush from retail therapy, try writing an actual letter to someone – anyone – you love, admire or appreciate. You don't need to know them personally. Writing down someone's positive characteristics, or their positive impact on your life, and sharing it with them (if you want to), is a free and scientifically proven way to boost your happiness levels.[1]

READ MORE BOOKS

Reading expands your mind and lowers your stress levels. Studies show that reading is even better at chilling you out than having a cup of tea, listening to music or going for a walk. Just six minutes of reading can zap your stress levels by more than two thirds.[2] So, join your local library or get down to your nearest bookshop and get lost. In a good way.

WE TIME

Social connections and support networks play an important role in our health and well-being – and are especially valuable to people actively engaged in changing the world for the better. Serious issues need serious antidotes, and what better medicine for overwhelm than a dance-off or some meaningful volunteering. When you're next in need of some communal capers, try some of these out on your existing friends...or use them as a way to make some new ones.

HAVE A GATHERING

As well as heating up your home in the winter (think of all that extra body heat), having friends over and sharing a meal is a great way to boost your happiness levels. Being with people face to face beats them liking your posts any day and, if you add some fun games into the mix, you'll be left wondering how you could have ever thought spending the afternoon at the mall was 'fun'.

DANCE

At home alone or with a local dance group, dancing is a brilliant way to connect with your body, release tension and live longer. Next time you feel unable to find the good in the world, give it up to dance and see what happens next.

DECLUTTER AND HAVE A GARAGE SALE

Letting go of stuff you no longer need is therapeutic; it frees up space, energy and resources. Selling things on eBay or at a car-boot sale (or yard sale, if you're in the US) is a great way to generate some cash and give a second life (or third, if it was second-hand when you acquired it) to your collection of knick-knacks.

WALK A DOG

If you don't already have a dog in your life, then how about borrowing someone else's doggy and taking them out for

a walk? It's a winner in so many ways. You get free exercise and furry four-legged friendship, you get to help someone out (most dog walkers could do with a day off from time to time) and you'll undoubtedly get a load of social interaction in green spaces near you. Borrowmydoggy.com is an awesome example of making it safe and simple to find dogs near you, if you live in the UK. Or you could contact a charity, such as the Cinnamon Trust, who look after pets of the elderly, frail and terminally ill, as they are always looking for dog walkers.

PLAY GAMES

Outdoor games if the weather's good, board games and other fun stuff inside when it's not. Most of us forget how good it feels to play with our family and friends. Games can be silly (my favourite kind) or strategic (my least favourite) and should definitely not only be for dusting off at Christmas. Find some inspiration online or get some second-hand games from a local charity shop, organize a games night and let the good times roll.

DO SOME VOLUNTEERING IN YOUR COMMUNITY

Doing good makes us feel good and feeling good ripples out into the world through what we do, say and create. If you've got enough food to eat, enough money to pay your bills and good enough health to get out of bed each day then you'd make a great volunteer. Notice I didn't say 'enough time'. Most of us think we don't have that. But I bet you could find one Saturday a month, or a year, to get involved with something that could use your energy. Think of what you'd like to change in the world and give a little bit of yourself to it.

MORE WAYS TO SAVE THE WORLD WHEN YOU PLAY

FORAGE FOR YOUR FOOD: rummaging around in nature's larder for ingredients is a fun thing to do. Fruit, herbs, mushrooms and leaves can be used to make preserves, jellies, medicines and meals, but you'll need to learn what's hot and what's not from an expert. Just make sure foraging is allowed where you're headed and only take what you need – leave plenty for wildlife and other foragers!

GO RETRO WITH YOUR BOARD GAMES: finding vintage board games from online auction sites or charity shops is a great way to relive some childhood memories while having fun for free.

MEDITATE: some might argue this isn't exactly play, but meditation can unlock some pretty interesting doorways that are there for you to explore.

DO A BEACHCLEAN OR LITTER PICK: beachcleaning is actually FUN! Especially when done with a group of likeminded people on an organized clean-up. Some beachcleans are part of citizen-science programmes too, so your findings are recorded and provide valuable data on the plastic pollution issue.

GO GEOCACHING: there are millions of geocached hidden treasures around the world, in towns and the countryside, so there's quite likely to be one near you. Geocaching is essentially the world's biggest treasure hunt where you use a GPS or phone to hide and seek containers. See the geocaching website or app to learn more.

LEARN TO COOK A NEW DISH OR BAKE A CAKE: using organic and local ingredients if possible, add something yummy to your repertoire. Try looking on Instagram for vegan and raw healthy 'bake' ideas.

DO A MASSAGE SWAP: our bodies love touch, so team up with a partner or friend, get the coconut oil out and let your stresses melt away.

TRY 'SCREEN-FREE SUNDAYS': going screen-free for a day each week (Sundays seem to work best as they're usually work-free and slower-paced than Saturdays) is an awesome way to use less electricity and enjoy the space it creates. And, yes, that includes your phone.

PLANT A BEEBOMB: terrifyingly, in the UK, 97 per cent of natural bee and butterfly habitat has been lost since the Second World War[3]. Beebombs, muddy balls full of bee-friendly plant seeds, are a happy antidote. Get them as Christmas presents, plant them wherever you can and wait for spring to arrive.

ACTUALLY BECOME A TREE-HUGGER: walking among trees and, even, making like a koala and throwing our arms around them, is reported to create a natural high and enhance wellbeing. In Japan, it's widely practiced as 'forest bathing' and people don't think you're weird. Really, who cares if they do anyway. You're just giving them permission to do it too.

CONNECT WITH THE COSMOS: climb a hill to see the sun rise or set, make sure you don't miss the next meteor shower, and/or go stargazing and marvel at the free show the universe puts on for us every day. Reminding ourselves that we're passengers on Earth as it travels at 107,000 kmh (66,600 mph) around the sun never hurt anyone.

JOIN A 'MEETUP': there are people in your local area meeting up to share their passion for something. See what interests you and go join them, for free.

WHEN HAVING A PARTY: go for real crockery, or at least bamboo or leaf plates that can be composted, get vintage tablecloths and Boho mismatching china plates from charity shops, paper straws, cotton bunting, paper party bags with homemade going-home presents for children, organic local drinks and water served in refillable glass jugs with lemon slices or mint. Parties are a great opportunity to lead by example without having to preach to anyone!

DURING THE FESTIVE SEASON: make your own wreaths with woodland leaves and berries; make decorations from dried orange slices, twiggy stars, pine cones or cinnamon sticks.

Save the World When You…Vote

Voting is one of the single most powerful things we can do to influence how our country and communities are run. Most of us are realizing that the power is in our hands – in the last few years more Brits have registered to vote than ever before, with over a million young people signing up ahead of the 2017 general election.[1] 2018 was a record-breaking year for National Voter Registration Day in the US, with more than 800,000 people registering to vote in a single day![2] And this is important, as, despite the power we have as individuals to save the planet, our political leaders have a whole lot more. In the 2016 US presidential election, just 63 million out of 250 million eligible voters backed Donald Trump. 73.5 million actively voted for another candidate, which included more than 65.8 million who cast a vote for Hillary Clinton.[3] Yet 45 per cent of adults who live in the US didn't vote – a huge kick in the geo-thermals for Planet Earth. So, if you haven't registered or voted previously and you care about the environment, voting offers an important opportunity to make a difference, especially if you're voting for a party with stellar environmental credentials, like the Green Party.

But let's not pretend that the reason so many people aren't voting is because they don't see the point. Modern-day democracy generally means that we vote for a self-selecting person, representing a party we don't really trust and, if the one we voted for loses, we spend the next four or five years feeling as though we have no say. It's no wonder lots of us don't bother.

REIMAGINING DEMOCRACY

Around the world, creative citizens are coming together to change our political systems for the better. Which is good news for the planet and crucial if we want to have an active role in the way decisions are made. People just like you and me are busy reimagining society and formulating ideas for policy...and they're getting elected.

A BETTER WAY THAT'S WAY MORE DEMOCRATIC

In Denmark, Alternativet (The Alternative) is a political party that is a more deliberative form of democracy. Frustrated with the Danish Parliament, Uffe Elbæk decided to do something different. In 2013 he formed a party that, instead of aligning with left- or right-wing ideology, is motivated by the values of courage, empathy, openness, generosity, humility and humour, and crowdsources its political programme. I'd vote for that in a flash. And, contrary to what his critics thought, in 2015 lots of people did just that. Instead of failing, Alternativet surprised everyone by gaining 5 per cent of the vote, which meant nine representatives in the Danish parliament.[4] As if I needed another reason to love Denmark.

Inclusive, citizen-led systems work, and result in more people voting on more issues. The people of Switzerland get to vote in over ten referendums a year on issues that affect them. The city of Porto Alegre in Brazil operates a participatory budget system whereby citizens are involved in identifying spending priorities and vote on which ones to implement. Residents of Reykjavík, the Icelandic capital, are enjoying better amenities and quality of life thanks to their participatory approach in which they guide their parliament's decisions: people simply propose an idea for improving the city or how its infrastructure budget should be spent, and anyone can vote for or against it.[5]

AN INSTRUCTION GUIDE TO REAL DEMOCRACY

In order to arrive at decisions that genuinely meet the needs of life on Earth, we may need to disrupt our dysfunctional political systems. And that can be a lot more fun – and easier – than it sounds. An example of this is Flatpack Democracy, a powerful and empowering approach that even comes with an instruction manual (of the same name) written by its founder, Peter Macfadyen, on how to take back political power from the elite few and put it back into the hands of the many. Founded in a small, but perfectly formed, market town in rural England, Flatpack Democracy put Frome, on the global map as an example of how, to paraphrase the American anthropologist Margaret Mead, a few thoughtful, committed citizens have changed the world.

During the local elections that accompanied the UK's General Election in 2015, the people of Frome rejected the parties they had normally supported and instead voted for Independents for Frome – citizens like you and me who'd been selected by an independent group of people. Independents for Frome took 10 out of 17 seats in 2011 and embarked on a programme of 'radical reformation'. Locals approved so much, they went on to elect 17 out of 17 seats on Frome's Town Council in the 2015 elections.[6]

What started as a small-scale political revolution is spreading across the world, creating a bottom-up approach in a top-down system, while making politics relevant, effective and fun in the process. People are having a greater say in how their communities are run and are able to give issues such as green spaces and air pollution a real voice. Flatpack Democracy breathes life into the politics of the future, doing away with the stuffy, disconnected politics of the past.

MEETING A MACHINE WITH A MOVEMENT

Hope for a fairer, more representative world comes also in the form of Big Organizing – an exciting people-powered development in politics over the past few years. Developed by Bernie Sanders's campaign strategists, Becky Bond and Zack Exley, it was the secret sauce (though not secret anymore as they wrote a very good book about it) behind the hugely successful – in terms of its rapid scaling-up – Sanders campaign for the democratic presidential nomination in 2015–16.

Essentially, Bond and Exley created a social movement. A massive volunteer-driven campaign that was built on supporter enthusiasm, trust and empowering individuals to make decisions as well as donations. Instead of going down the traditional route of funding the campaign by accepting huge sums given by big business (which in turn would make Sanders indebted to those industries) it was funded by an army of small donors, proving that a people-powered approach can compete with even the wealthiest individuals and corporations.

All this is good news, as not only did Big Organizing reveal the enormous number of American citizens willing to fight for radical change – such as slashing US carbon pollution by creating a carbon tax – it set a precedent for how to change the world, at scale.

Knowing that this approach works, it can be refined and replicated, as was demonstrated by then 28-year-old Alexandria Ocasio-Cortez in 2018. Having worked on the Sanders campaign, the daughter of a Puerto Rican mother and a Bronx-born father from a working-class community, rose up to beat one of the Democratic party's most senior congressman in a remarkable election victory in New York. All the more exciting is the fact that her opponent, Joe Crowley, had raised over US$3 million for his campaign, which was ten times the amount raised by Ocasio-Cortez.[7] She'd shunned corporate donations, as had Sanders in 2016, and proven that this time it was about the people, not the money.

Ocasio-Cortez's election brings fresh hope to environmental issues in the US too, her 'Green New Deal' idea has, at the time of writing, been championed by 15 Democratic House lawmakers as a way to create a select committee to focus on transitioning the country towards 100 per cent renewable energy.

THE FUTURE IS GREEN

Whether or not you feel the call to reclaim your power as a citizen, you can still do a lot of good with your vote when election time comes around by voting Green. The Green Party of England and Wales has made a commitment – the only party to do so – to a 'life based on democracy and justice within the planet's limits', and hundreds of Greens who've made the same commitment around the United States hold positions on the municipal level, including on school boards, city councils and as mayors.

The scale of the environmental issues we're facing demands that we put our environment at the heart of everything we do, not treat it as a cheap add-on at the end. That's why Greens have embedded the living 'within the planet's limits' ethos as one of their core values. This means their objectives don't depend on who the current leader is or in which direction the nation's political winds are blowing.

DEMAND CREATES SUPPLY

'But my local Green candidate has no chance of winning here!', you might say. Does that mean voting for them is a waste? Not at all, and here's why. In the UK, Greens in places such as Bristol and Brighton have gone from 'can't win here' to being elected, to making a difference. So, for example, in 1997 Greens won 2.7 per cent of the vote in the Brighton Pavilion ward, trebled it in 2001, doubled that in 2005 and, in 2010, they won the seat with 31.3 per cent – and their vote share has continued to grow since. Imagine if the good people of Brighton had thought, 'Nah, Greens can't win here' back in the 2000s – they wouldn't have had, in my opinion, one of the most badass MPs, in Caroline Lucas, since 2010. Demand for Greens is growing in the United States too – because of the performance of their presidential candidates in 2016, state Green parties have maintained ballot access (the right to nominate candidates) in 21 states, meaning more post-election ballot lines (names on the ballot sheet) than they've had in over a decade.[8]

REFORM FOR A FAIRER SYSTEM

But it isn't that easy to get Greens elected. For almost 100 years, politicians who most of us didn't vote for – and don't agree with – have been governing the UK and the US. How so? Well, over 60 countries around the world use the 'first past the post' electoral system, or FPTP for short. In the UK, for example, this means that, to become an MP, a candidate only has to win more votes than any rival in their area, so the governing party need not necessarily win a majority of votes cast. During an election, millions of people can vote for Green Party candidates and only end up with a single MP representing them in parliament, whereas a few hundred thousand voters for another party can get ten times as many MPs.

We don't just have to accept the status quo. By adopting a 'Proportional Representation' (PR) voting system, successful parties gain seats in direct proportion to the number of votes they get in an election. Though the system of PR is not itself without its flaws, it is hard to argue that it isn't in most senses fairer than FPTP. If you're in the UK and in favour of PR, sign up to the Make Votes Matter campaign. It's free and is a powerful step towards saving the world.

BE THE CHANGE – STAND FOR ELECTION

Other than eating a lot less meat and taking fewer flights, the best way to save the world is to get out there and change it. Yep, that means rolling our sleeves up and getting political. Which, as we've seen in the examples of Flatpack Democracy and Big Organizing, can be a lot more interesting than it sounds. For some, the word 'politics' can conjure images of stuffy, out-of-touch, self-interested so-called leaders. Politicians aren't always the kind of people we look at and think, 'Yes! You've got my best interests and the interests of the planet at heart!' But check out the world of local government and things start to get more interesting. There is a groundswell of forward-thinking elected officials from a range of backgrounds, some not affiliated to any parties at all, out there making waves in our communities. And it could include you.

LOCAL LEADERSHIP

When you're elected on the municipal level, on local or city councils, you get to make cool stuff happen. Real, lasting change that improves the quality of life for people living in your area and, if you're so inclined, changes that put a serious dent in carbon emissions, use of pesticides and air pollution. Being elected gives you power, a power which you then need to use – no matter how small – constructively and creatively. And some days, against all odds, that power pays off. In 2018, Green Party city councillor Carla Denyer put forward a motion to the city council for Bristol to be carbon neutral by 2030 instead of 2050, in line with the IPCC's report on climate change stating we had 12 years in which to avoid catastrophic climate breakdown. The council voted unanimously in favour of the commitment. How's that for saving the world?

Getting elected won't cost you anything either, so it definitely counts as saving the world for free too. In fact, at the higher municipal levels, you get paid an allowance for your expenses when you get elected. So that's a win for Planet Earth, a win for your local community and a win for you too.

MORE WAYS TO SAVE THE WORLD WHEN YOU VOTE

SUPPORT WOMEN TO STAND FOR ELECTION: only 20 per cent of all congressional seats in the US are held by women[9] and only 32 per cent of MPs in the UK are women.[10]

FIND OUT WHICH PARTY BEST MATCHES YOUR VALUES: take the Vote for Policies survey (UK) or take the quiz on iSideWith (US).

CAMPAIGN FOR A SYSTEM OF PROPORTIONAL REPRESENTATION: as an alternative to 'first past the post', if appropriate for your country.

SERVE YOUR COMMUNITY OR COUNTRY IN OFFICE: get more involved with politics and take a look at beacouncillor.co.uk (UK) or runforoffice.org (US).

Maximize Your Impact

You are not alone. Although one person *can* change the world, it's a lot more fun to do it together. And usually – but not always – the person who looks as though they're changing the world on their own has a great team behind them, or a supportive partner or family who keep them charged up. It's a team effort. And how good would it be to have the government on your team too? Or hundreds of thousands of people who feel the same way you do? Well, you can. In this section, we'll explore some ways we can all amplify the impact of our actions.

SHARE AND SHARE AND LIKE

Whenever you make a switch from something you were doing and the green alternative feels better, smells better or tastes better, make sure you tell all your friends. Social media has its downside but, as a tool for social change, it's incredibly effective – never doubt the power of sharing a petition, post or film on social media. Whether it's a quick video recorded on your phone, a rushed photo or a work of art, get it out there. Share your solutions.

A NEW NARRATIVE

Sharing something personal, that's made a difference to your life, can be even more influential, as we humans are more likely to act if the message comes from people we love and trust.

Whether it's a super-simple switch from a plastic scouring pad to a natural loofah, or something bigger, such as joining a climate march of hundreds of thousands of protesters, telling your story is how we build a movement.

The more of us telling green, solution-focused stories, the faster we shift the focus from what's wrong with the world to what's right. And, when those 'right' things are seen as normal, the whole world will want to do them.

THE POWER OF PETITIONS

Online petitions are a fantastic tool for change-making, and they're available to anyone who has an internet connection. Handing in a physical petition can be an important media event, too, if you're running a campaign. A petition's success can depend on a number of factors, from how well-written the 'ask' is, to the photo that depicts the issue, or how well-connected the person that starts it is. But the fact remains that online petitions offer a route to lasting change and, sometimes, in a matter of weeks. Change.org, Avaaz, Sum of Us and 38 Degrees are great examples of people-powered petition sites that are making waves. My organization's 'Switch the Stick' petition on the 38 Degrees platform asked UK retailers to stop making the stems of cotton buds (Q-tips) out of plastic and switch to paper instead. 155,000 signatures later and all nine major retailers had agreed to make the change, a move which stopped over 400 tonnes of single-use plastic at source, every year in the UK.

USING CONSUMERISM AS A TOOL FOR CHANGE

The above example demonstrates how we can use consumerism for good. Our supermarkets and retailers need us; without us buying their products, they can't exist. Never underestimate how important you are to them and how, in fact, you have all the power. If you want refillable groceries in your supermarkets, tell them and get all your friends to tell them. If you want them to stop using palm oil, make sure they know you're not going to spend your money in their stores until they do. If there are a few hundred thousand of you saying the same thing, they will have to listen or risk losing customers.

NAMING, SHAMING AND GIVING THANKS

I'm all for positivity and sharing solutions, but when it comes to corporations not taking responsibility for the mess they're making, a bit of naming and shaming doesn't hurt. If your favourite makeup brand has unrecyclable packaging, make a song and dance about it by writing to them or calling them out on social media. Or you could switch to a more ethical brand and champion them publicly for looking after people and planet as well as their profits. In addition to encouraging them to keep doing it, thanking companies that are doing the right thing has the added bonus of promoting their product to other people who might benefit from your findings. Your voice can make a huge difference, so use it often.

GETTING YOUR GOVERNMENT TO LISTEN

This one's harder, as corporate lobbyists have a great deal of access to, and influence over, many of our elected representatives. Frustratingly, it's often the case that their interests carry more weight than those of the average citizen. Corporate lobbyists aside, ultimately politicians want to stay in power and can be influenced by their supporters, especially if they think it can win them even more votes. Get to know the people that represent you in Parliament or Congress and at a local level, too, in your town, city and county councils. Go to your representatives' surgeries, write them letters and engage them in the issues you care about. And make sure you're registered to vote.

GETTING CRAFTY

Activism comes in many forms. A particularly beautiful one is craftivism, in which people use different forms of craft work to bring about change in the world. This is perfect for people who prefer to 'provoke not preach' and want to take a quieter, more reflective path to protest. UK-based Sarah Corbett, founder of the Craftivist Collective, was behind the successful 'Don't Blow It' campaign that included bespoke, handstitched handkerchiefs being given to 14 board members of a large UK retail company. The campaign resulted in 50,000 of the retailer's employees having their pay increased above the minimum wage and to the current UK Living Wage calculations. A gentle, playful approach that's clearly not to be sniffed at.

IN THE BELLY OF THE BEAST

Change can come from within, of course, and I don't just mean in terms of meditating more and feeling less anxious. In this case, I'm talking about changing the system, or the company you work for, from the inside out. I know some brilliant examples of people who, when they switched on to sustainability, managed to get huge changes happening in their workplaces, from online retailers, to banks and, even, a major restaurant chain. After all, you've got more than a foot in the door, you've got a desk, the ability to talk to people in different departments and, in some cases, access to the supply chain. Done in the right way, you can use your power and influence to make change happen. Whether it's bringing in speakers to open people's minds on their lunch breaks, getting your canteen to choose local organic food, switching disposable cups and packaging for reusables, or getting the company to become a certified B-Corp – businesses that balance purpose and profit – you can be the activist within. And, when your employer has thousands of employees and a long supply chain, those small changes can have a massive impact.

DO WHAT YOU LOVE AND LOVE WHAT YOU DO

The best way to maximize your impact is, ultimately, to be yourself. Bending yourself out of shape to do volunteer work that doesn't suit you isn't going to light you up. But doing something you love, or that makes you shine, will be irresistible. Build on your strengths, do what comes naturally to you and be who you are. Get clear on your motivation for changing the world and tell the story in a way that's unique to you – knitting or fundraising, scientific research or filmmaking – wherever your talents lie, you can make them count by applying them to saving the world.

WATCHLIST

A good environmental documentary can be a great way to fire up our eco-cylinders and inspire us to take action. In my case, even watching the trailer for *Albatross* (listed below) was enough to change my life and propel me to have a positive impact. Sometimes, nothing beats seeing environmental issues with your own eyes and, while it may not be first-hand, it's often enough to wake us up to the reality of our actions. That said, all the films listed here are from the storytellers' perspective, so it's never a bad idea to do some of your own research and keep an open mind. And thank you to my fabulous Facebook community for the suggestions below.

Grab yourself some loose popcorn kernels from your local scoop shop, invite some friends over and get eye-popping at this lot. You can find links to watch most of them for free at www.nataliefee.com/savetheworld.

FOOD AND FASHION:

Earthlings, 2005

The End of the Line, 2009

Forks over Knives, 2011

Cowspiracy, 2014

The True Cost, 2015

Seed: The Untold Story, 2016

OCEANS:

Blackfish, 2013

Mission Blue, 2014

Plastic China, 2016

A Plastic Ocean, 2016

Albatross, 2017

Blue, 2017

Chasing Coral, 2017

CLIMATE:

An Inconvenient Truth, 2006

Gasland, 2010

Gasland Part II, 2013

This Changes Everything, 2015

Before the Flood, 2016

An Inconvenient Sequel, 2017

ACTIVISM:

Just Do It: A Tale of Modern-Day Outlaws, 2011

Elemental, 2012

Tomorrow [originally *Demain*], 2015

CITIZENSHIP, POLITICS AND ECONOMICS:

The Story of Stuff, 2007

The Four Horsemen, 2012

All Hands On series, episode 1 'When Citizens Assemble', 2017

PEOPLE AND MOVEMENTS TO FOLLOW

charleseisenstein.net

democracycollaborative.org

eradicatingecocide.com

extinctionrebellion.org

gofossilfree.org

monbiot.com

naomiklein.org

twitter.com/GretaThunberg

participatoryeconomics.info

positive.news

positivemoney.org

ronfinley.com

storyofstuff.org

theecologist.org

transitionnetwork.org

treesisters.org

wearestillin.com

350.org

A BETTER WAY THAT'S WAY MORE FUN

While researching this book, I did a lot of crying. I cried over lives lost protecting forests from the impact of cattle farming in Brazil, over the lives lost making our clothes in Cambodia and over the treatment of the people of the Marshall Islands by the US military over the past 50 years: to be fair, I may have ended up down a few rabbit holes. I also raged over the way animals are treated and the injustice of our economic and political systems. And I had dreams in which I was talking to a dolphin behind glass about overfishing, and another in which I was proudly showing people at a conference the label in my knickers which stated who'd made them, which factory they worked in and how much they were paid. So, you could say, the experience has been immersive.

But beyond the grief, rage and strange dreams, I've mostly felt very grateful. To be in a position to write this book is a privilege, and to have you read it is an honour. And, on top of that, I am truly, madly and deeply in love with this planet. (Not to the point of considering myself an eco-sexual, but still, I can understand the appeal.) Aside from food and shelter, it's the beauty of it all that gets me, from the ant colonies to the mountains, the babbling brooks to the sea. The relationships we can form with places, whether a tree or a whole forest, feed us in more ways than the physical, many of which we've yet to comprehend. And, like any relationship that I care about,

I want to look after it, cherish it, celebrate and nurture it, as if my life depended on it. Which, I suppose, in this instance, it actually does.

The innovation that's constantly emerging in response to the environmental crisis we face is extraordinarily exciting. The depth of connection we can discover on our doorsteps when we get more involved with our local communities is heart-warming. And the realization that we, both as individuals and collectively, can change the world for the better is a game changer.

And so, as we come to the end of our time together, for now at least, I invite you to take the bits you loved in the book, to embody and embrace them and share them with anyone you think will listen.

Here's to clean air, pure water, lush forests, healthy soil, energy from the sun, abundant wildlife and people we love to appreciate it all with.

ENDNOTES

Links for all of the sources cited in this book can be found by clicking on the 'Downloads & extras' tab at: www.laurenceking.com/product/how-to-save-the-world-for-free/

INDEX

AUTHOR'S ACKNOWLEDGEMENTS

My thanks, first and foremost, go to Zara Larcombe, at Laurence King Publishing, for somehow tuning in to the fact that I'd written a book proposal for this very book two weeks before she contacted me to ask if I'd like to write it. And for recognizing that the world needs it. My editor, Chelsea Edwards, has been everything I could have hoped for – clued-up, switched on and brilliant at telling me when things had gone a bit 'woo'. Sally Beare, my wing-woman, thank you for helping me stay on track, keep it real and get my head around the details. To Carissa Tanton – thank you for the gorgeous cover and illustrations throughout this book!

Most of what I've written about is based on the work of scientists and environmentalists around the world, who've been researching and reporting on these issues for decades. I'm hugely grateful to them, as well as the journalists who've written countless articles on their findings. This book would not be here without them and I hope it serves to further the impact of their work.

My network of fire-starters and changemakers have been endlessly helpful. I would especially like to thank Steve Hyndside, Molly Scott-Cato, Bevis Watts, Jo Morley, Suzi Martineau, Peter Macfadyen, Jules Peck, Tony Greenham, Fionn Travers-Smith and Steve Clark for their expert insight and valuable input. To the fantastic team at City to Sea, thank you for carrying our work into the world and, particularly, to Becci, Gus and Michelle for holding the plastic-free fort while I wrote this book.

My heartfelt thanks also go to: Sophie & Charlott for tending the fire of my women's work; Nick and Craig at The Wave for the front-row seats; Adam Hall at Surfdome for showing what's possible in a big business; Martin Dorey from #2minutebeachclean for leading the way; Hywel George for inspiring a lifelong love of cycling and for the silly cycling joke; Chris Jordan for *Albatross*; Brian and Tommy at Dancing Fox for helping me find a better way that's way more fun; Daniel Dobbie for his world-changing events; the NGOs on the front line of all the issues I've written about; my much-loved Facebook community for tips and documentary recommendations; Stephen and Lynda Kane for showing me the 'Way' and giving me the tools, and love, to live life to the full.

To my family – thank you for all your encouragement and support, especially my Mum, Frances, who helped me find my love of words. And to my partner, Angus, for being the best researcher a woman could wish for and for helping me stay sane, warm, fed and happy while writing this book, despite sitting me down to watch films that kept making me cry. And, finally, to my son, Elliot, thanks for putting up with my WiFi-free, plastic-free, free-range, switch-everything-off-at-the-plugs parenting.

ABOUT THE AUTHOR

Natalie Fee is an award-winning environmentalist, author, speaker and founder of City to Sea, a UK-based organization running campaigns to stop plastic pollution at source. In 2018, Natalie was listed as one of the UK's '50 New Radicals' by *The Observer* newspaper and NESTA, the innovation charity, and, in the same year, the University of the West of England awarded her an honorary doctorate of science in recognition of her campaign work. She was also named as Bristol 24/7 website's Woman of the Year. In 2017, Natalie won the Sheila McKechnie Foundation's Award for Environmental Justice, for City to Sea's #SwitchtheStick campaign, and gave her first TEDx talk, 'Why Plastic Pollution is Personal'. She can be found sharing more world-saving tips on Instagram as nataliefee_ and on Twitter as nataliefee. She invites you to connect and share your trials and triumphs with her using the hashtag #HowToSaveTheWorldForFree.

For more information on Natalie's speaking events, workshops and campaigns visit:

nataliefee.com/savetheworld
citytosea.org.uk

10 per cent of the author's proceeds from book sales will go to support City to Sea's campaigns.